Combinatorics on Words

Christoffel Words and Repetitions in Words

Volume 27

CRM MONOGRAPH SERIES

Centre de Recherches Mathématiques
Montréal

Combinatorics on Words

Christoffel Words and Repetitions in Words

Jean Berstel
Aaron Lauve
Christophe Reutenauer
Franco V. Saliola

American Mathematical Society
Providence, Rhode Island USA

The production of this volume was supported in part by the Fonds Québécois de la Recherche sur la Nature et les Technologies (FQRNT) and the Natural Sciences and Engineering Research Council of Canada (NSERC).

2000 *Mathematics Subject Classification.* Primary 68R15;
Secondary 37B10, 11J70, 68W40.

For additional information and updates on this book, visit
www.ams.org/bookpages/crmm-27

Library of Congress Cataloging-in-Publication Data

Combinatorics on words : Christoffel words and repetitions in words / Jean Berstel ... [et al.].
p. cm. — (CRM monograph series ; v. 27)
Includes bibliographical references and index.
ISBN 978-0-8218-4480-9 (alk. paper)
1. Combinatorial analysis. 2. Word problems (Mathematics) I. Berstel, Jean, 1941–

QA164.C6638 2008
511′.6–dc22 2008036669

Printed in the United States of America.

∞ The paper used in this book is acid-free and falls within the guidelines
established to ensure permanence and durability.
This volume was submitted to the American Mathematical Society
in camera ready form by the Centre de Recherches Mathématiques.
Visit the AMS home page at http://www.ams.org/

10 9 8 7 6 5 4 3 2 1 14 13 12 11 10 09

Ce livre est dédié à la mémoire de Pierre Leroux (1942–2008)

Contents

Preface

This book grew out of two series of five two-hour lectures, given by Jean Berstel and Christophe Reutenauer in March 2007. The lectures were delivered during the school on "Combinatorics on Words" organized by Srečko Brlek, Christophe Reutenauer and Bruce Sagan that took part within the theme semester on *Recent Advances in Combinatorics on Words* at the Centre de recherches mathématiques (CRM), Montréal, Canada.

Notes for the lectures were written down by Aaron Lauve and Franco Saliola. They have augmented their notes with several topics and have added more than 100 exercises. There has been a lot of work in adding bibliographic references and a detailed index.

The text is divided into two parts. Part I, based on the lectures given by Christophe Reutenauer, is a comprehensive and self-contained presentation of the current state of the art in Christoffel words. These are finitary versions of Sturmian sequences. It presents relationships between Christoffel words and topics in discrete geometry, group theory, and number theory. Part I concludes with a new exposition of the theory of Markoff numbers.

Part II, based on the lectures by Jean Berstel, starts with a systematic exposition of the numerous properties, applications, and interpretations of the famous Thue – Morse word. It then presents work related to Thue's construction of a square-free word, followed by a detailed exposition of a linear-time algorithm for finding squares in words. This part concludes with a brief glimpse of several additional problems with origins in the work of Thue.

Acknowledgements. We gratefully acknowledge the generosity of Amy Glen and Gwénaël Richomme, who agreed to read a preliminary version of this text. Implementation of their numerous comments improved the quality of the text tremendously. We also thank Anouk Bergeron-Brlek for lending us a third set of notes and Lise Tourigny through whom all things are possible. Finally, we would like to thank the CRM and François Bergeron, the principal organizer of the CRM theme semester, for providing an excellent scientific program and working environment during the semester as well as support throughout the preparation of this text.

Typesetting. The book was typeset with the LaTeX document preparation system together with the following LaTeX packages:

`algorithm2e`	`color`	`multicol`	`stmaryrd`
`amsrefs`	`ednotes`	`paralist`	`subfigure`
`amssymb`	`gastex`	`pstricks`	`upref`
`array`	`graphicx`	`pst-poly`	`xy`
`bbm`	`mathtools`	`pst-tree`	

and Will Robertson's LaTeX code for typesetting magic squares [Rob2005].

Notation

We gather in one place the notational conventions shared by the two parts. The reader may also consult the subject index to locate the major occurrences within the text of most of the symbols and bold words below.

Let $\mathbb{N}$ denote the set of nonnegative integers. If a, b and n are integers, then the notation $a \equiv b \bmod n$ shall mean that $a - b$ is divisible by n. Equivalently, $a \equiv b \bmod n$ if and only if a and b have the same remainder upon division by n.

Let A denote a finite set of symbols. The elements of A are called **letters** and the set A is called an **alphabet**. A **word** over an alphabet A is an element of the free monoid A^* generated by A. The identity element ϵ of A^* is called the **empty word**. Given a word $w \in A^*$, the **square** of w is the monoid product $w^2 = ww$ in A^*. Higher powers of w are defined analogously. We frequently take A to be a subset of the nonnegative integers $\mathbb{N}$. The reader is cautioned to read 101 not as "one hundred and one" but as "$1 \cdot 0 \cdot 1$," an element of $\{0,1\}^3$.

If $w \in A^*$, then there exists a unique integer $r \geq 0$ and unique letters $a_1, a_2, \ldots, a_r \in A$ such that $w = a_1 a_2 \cdots a_r$; the number r is called the **length** of w and denoted by $|w|$. A positive integer p is a **period** of w if $a_i = a_{i+p}$ for all $1 \leq i \leq |w| - p$. (Note that if $p \geq |w|$, then p is a period of w.) If $w \in A^*$ and $a \in A$, then $|\boldsymbol{w}|_{\boldsymbol{a}}$ denotes the number of occurrences of the letter a in the word w so that

$$|w| = \sum_{a \in A} |w|_a.$$

If $w = a_1 a_2 \cdots a_r$, where $a_1, a_2, \ldots, a_r \in A$, then the **reversal** of w is the word

$$\widetilde{w} = a_r \cdots a_2 a_1.$$

We say w is a **palindrome** if $w = \widetilde{w}$.

An **infinite word** is a map from $\mathbb{N}$ to A, typically written in bold or as a sequence such as $\boldsymbol{w} = w(0)w(1)w(2)\cdots$ or $\boldsymbol{w} = w_0 w_1 w_2 \cdots$ (we freely pass between the two notations w_n and $w(n)$ in what follows). Any finite word m gives rise to a periodic infinite word denoted $\boldsymbol{m}^{\boldsymbol{\infty}}$, namely

$$m^\infty = mmm \cdots .$$

A **factorization** of a finite word w over A is a sequence $(w_1, w_2, \ldots, w_r)$ of words over A such that the relation $w = w_1 w_2 \cdots w_r$ holds in the monoid A^*. We sometimes write $w = (w_1, w_2, \ldots, w_r)$ to emphasize a particular factorization of w. Factorizations of infinite words are similarly defined (with w_r necessarily the only infinite word in the sequence). If w is a finite or infinite word over A and $w = uv$ for some (possibly empty) words u and v, then u is called a **prefix** of w and v is a **suffix** of w. Conversely, a **factor** of a finite or infinite word w is a finite word v such that $w = uvu'$ for some words u, u'; we say v is a **proper factor** if $v \neq \epsilon$ and $uu' \neq \epsilon$. Given two words $w, w' \in A^*$, we say that w is a **conjugate** of w' if there exists $u, v \in A^*$ such that $w = uv$ and $w' = vu$.

Let w be a finite or infinite word over an alphabet A and write $w = a_0a_1a_2\cdots$, where $a_0, a_1, a_2, \cdots \in A$. If v is is a factor of w, then

$$v = a_ia_{i+1}\cdots a_j \quad \text{for some} \quad 0 \leq i < j,$$

and $a_ia_{i+1}\cdots a_j$ is said to be an **occurrence** of v in w. (Specifically, an occurrence of v in w also includes information about where it appears in w; for the factor above, we say the **starting index** is i.) If u and v are words, then u is said to **contain** v if there is an occurrence of v in u.

Given two alphabets A, B, a **morphism** from A^* to B^* shall always mean a "morphism of monoids." That is, a set mapping $f\colon A^* \to B^*$ satisfying

$$f(uv) = f(u)f(v) \quad \text{for all } u, v \in A^*.$$

In particular, $f(\epsilon_{A^*}) = \epsilon_{B^*}$ since the empty word ϵ is the only element in a free monoid satisfying $w^2 = w$. The **identity morphism** on A^* is the morphism sending each $w \in A^*$ to itself. The **trivial morphism** from A^* to B^* is the morphism sending each $w \in A^*$ to ϵ_{B^*}.

Part I

Christoffel Words

The goal of Part I of the text is to present a comprehensive and self-contained account of the combinatorics of Christoffel words, named after the German mathematician and physicist Elwin B. Christoffel (1829 – 1900). Since their first appearance in the literature, arguably as early as 1771 in Jean Bernoulli's study of continued fractions [Ber1771], many relationships between Christoffel words and other areas of mathematics have been revealed. After laying out the current state of the art in Christoffel words, we close by recounting some of these relationships in the last four chapters.

CHAPTER 1

Christoffel Words

Although the theory of Christoffel words began to take shape in the late 1800s [Chr1875, Smi1876, Mar1879, Mar1880, Mar1881, Chr1887], the term was not introduced until 1990 by Jean Berstel [Ber1990]. By now there are numerous equivalent definitions and characterizations of Christoffel words. We choose as our working definition and point-of-view a geometric one: a Christoffel word is a "discretization" of a line segment in the plane by a path in the integer lattice $\mathbb{Z} \times \mathbb{Z}$ [OZ1981, Ber1990n, BL1993].

1. Geometric definition

Notation. If $a, b \in \mathbb{N}$, then a and b are said to be **relatively prime** if 1 is the only positive integer that divides both a and b. The notation $\boldsymbol{a \perp b}$ shall mean "a and b are relatively prime".

Suppose $a, b \in \mathbb{N}$ and $a \perp b$. The **lower Christoffel path** of slope b/a is the path[1] from $(0,0)$ to (a,b) in the integer lattice $\mathbb{Z} \times \mathbb{Z}$ that satisfies the following two conditions.

(i) The path lies below the line segment that begins at the origin and ends at (a,b).

(ii) The region in the plane enclosed by the path and the line segment contains no other points of $\mathbb{Z} \times \mathbb{Z}$ besides those of the path.

Upper Christoffel paths are defined analogously, using paths in $\mathbb{Z} \times \mathbb{Z}$ that lie above the line segment. See Figure 1 for examples. The unmodified term **Christoffel path** will always mean *lower* Christoffel path.

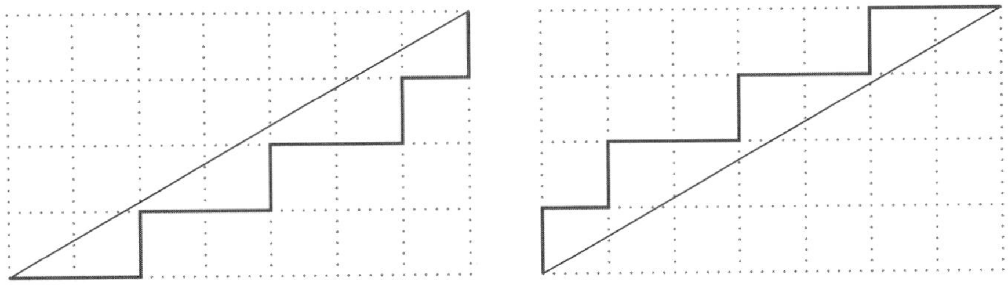

FIGURE 1. The lower and upper Christoffel paths of slope $4/7$.

[1] By a path in $\mathbb{Z} \times \mathbb{Z}$ from (a,b) to (c,d) we actually mean a continuous map $\alpha\colon [0,1] \to (\mathbb{Z} \times \mathbb{R}) \cup (\mathbb{R} \times \mathbb{Z})$ such that $\alpha(0) = (a,b)$ and $\alpha(1) = (c,d)$. Since such paths are essentially determined by the points of $\mathbb{Z} \times \mathbb{Z}$ that lie on the path, we identify such a path with a sequence of points in $\mathbb{Z} \times \mathbb{Z}$ with consecutive points of the sequence differing by $\vec{e}_1$ or $\vec{e}_2$, where $\vec{e}_1$ and $\vec{e}_2$ are the standard basis vectors of $\mathbb{R}^2$.

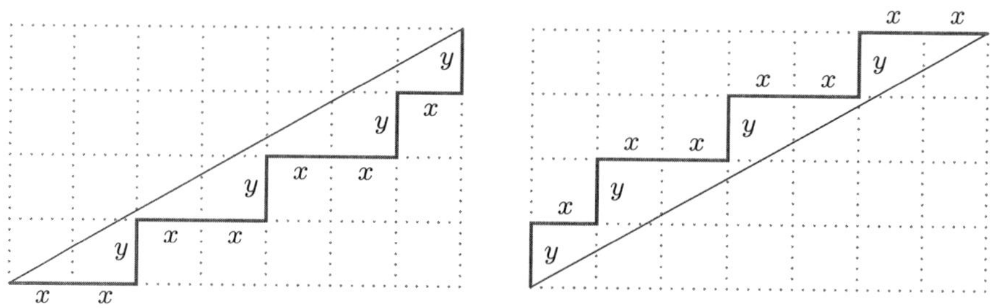

FIGURE 2. The lower and upper Christoffel words of slope $4/7$ are $xxyxxyxxyxy$ and $yxyxxyxxyxx$, respectively.

Since every step in a Christoffel path moves from a point $(i,j) \in \mathbb{Z} \times \mathbb{Z}$ to either the point $(i+1,j)$ or the point $(i,j+1)$, a Christoffel path of slope b/a determines a word $C(a,b)$ in the alphabet $\{x,y\}$ by encoding steps of the first type by the letter x and steps of the second type by the letter y. See Figure 2.

Definition 1.1. Let $a, b \in \mathbb{N}$. A word $w \in \{x,y\}^*$ is a **(lower) Christoffel word of slope** b/a if $a \perp b$ and $w = C(a,b)$. A Christoffel word is **trivial** if its length is at most 1, and is **nontrivial** otherwise. **Upper Christoffel words** are defined analogously.

Since every positive rational number can be expressed as b/a where $a \perp b$ in only one way, there is a unique lower Christoffel word of slope r for all positive rational numbers r.

Examples. The following are examples of Christoffel words.

(1) The Christoffel word of slope 0 is x, since $C(1,0) = x$.
(2) The Christoffel word of slope ∞ is y, since $C(0,1) = y$.
(3) The Christoffel word of slope 1 is xy, since $xy = C(1,1)$.
(4) The Christoffel word of slope $4/7$ is $xxyxxyxxyxy$ (see Figure 2).

Remarks.

(1) The empty word ϵ is *not* a Christoffel word since $0 \not\perp 0$. Therefore, x and y are the only trivial Christoffel words.

(2) The square or higher power of a Christoffel word is *not* a Christoffel word. Nor is a Christoffel word the power of a shorter word, that is, Christoffel words are **primitive** words. These statements follow from the observation that the number of occurrences of the letters x and y in the kth power of a word are both multiples of k (so they cannot be relatively prime if $k \geq 2$).

(3) Christoffel words have a natural generalization to infinite sequences: replace the defining line segment of slope b/a with an infinite ray of irrational slope before building the lattice path. The resulting right-infinite word is called a (characteristic) **Sturmian word**. See [Pyt2002], [Lot2002, Chapter 2] or [AS2003] for more information. While many of the references cited in what follows offer results at this level of generality, we restrict ourselves to Christoffel words here.

Exercise 1.1. Christoffel words are primitive words, as are all of their conjugates.

Exercise 1.2. Suppose a and b are nonnegative integers. Then $a \perp b$ if and only if the line segment from $(0,0)$ to (a,b) contains no integer points besides $(0,0)$ and (a,b).

Exercise 1.3. Suppose $a \perp b$. Prove that the region bounded by the segment from $(0,0)$ to (a,b) and the Christoffel path from $(0,0)$ to (a,b) has area $\frac{1}{2}(a+b-1)$. (*Hint*: Consider the region bounded by the upper and lower Christoffel words.)

Exercise 1.4. Suppose a and b are nonnegative integers. Then $a \perp b$ if and only if $a \perp (a+b)$.

Exercise 1.5 (Fibonacci word). Let $\phi^\vee$ denote the conjugate $(1-\sqrt{5})/2$ of the golden ratio. Using the Christoffel construction, compute the first 16 or so letters in the (Sturmian) word $\boldsymbol{s}$ corresponding to the ray of slope $-\phi^\vee$. The infinite word $\boldsymbol{f}$ satisfying $\boldsymbol{s} = x\boldsymbol{f}$ is called the **Fibonacci word**. (The word $\boldsymbol{f}$ is the "cutting sequence" of the ray of slope $-\phi^\vee$: it records the order in which the ray intersects the lines $x = i$ and $y = j$ for $i, j \in \mathbb{N}$.)

2. Cayley graph definition

We introduce an equivalent definition for the Christoffel word of slope b/a that will occasionally be useful. In fact, it is the definition originally used by Christoffel [Chr1875]. It amounts to reading edge-labellings of the Cayley graph of $\mathbb{Z}/(a+b)\mathbb{Z}$.

Definition 1.2. Suppose $a \perp b$ and $(a,b) \neq (0,1)$. The **label** of a point (i,j) on the (lower) Christoffel path of slope b/a is the number $(ib - ja)/a$. That is, the label of (i,j) is the vertical distance from the point (i,j) to the line segment from $(0,0)$ to (a,b).

The labels $1/7$ and $10/7$ from Figure 3 hold a special place in the theory. We return to them in Chapters 3 and 2, respectively. Exercise 1.8 gives an interesting fact about the label $3/7$ (or rather, the number 3).

Now, suppose w is a lower Christoffel word of slope b/a and suppose $(s/a, t/a)$ are two consecutive labels on the Christoffel path from $(0,0)$ to (a,b). Either $(s/a, t/a)$ represents a horizontal step (in which case $t = s + b$) or it represents a vertical step (in which case $t = s - a$). The following lemma summarizes these observations.

Lemma 1.3. *Suppose w is a lower Christoffel word of slope b/a and $a \perp b$. If s/a and t/a are two consecutive labels on the Christoffel path from $(0,0)$ to (a,b),*

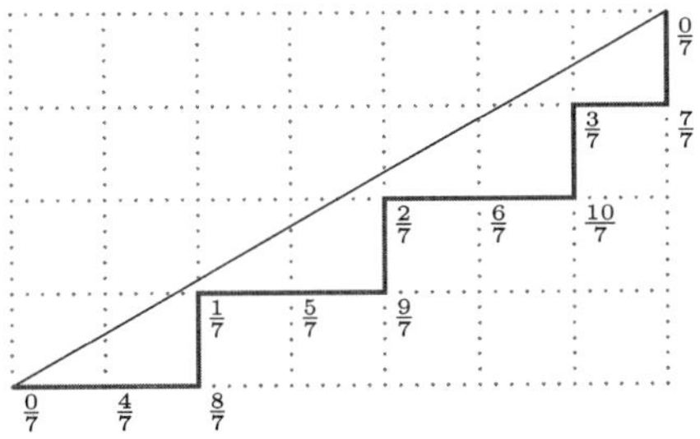

FIGURE 3. The labels of the points on the Christoffel path of slope 4/7.

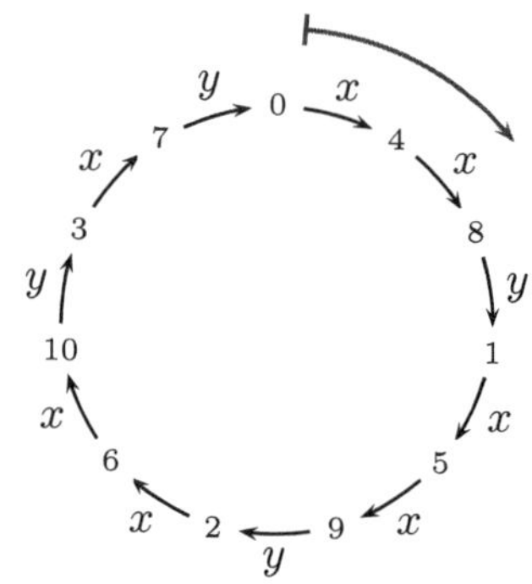

FIGURE 4. The Cayley graph of $\mathbb{Z}/(7+4)\mathbb{Z}$ with generator 4 and the associated Christoffel word.

then $t \equiv s+b \bmod (a+b)$. *Moreover,* t *takes as value each integer* $0, 1, 2, \ldots, a+b-1$ *exactly once as* $(s/a, t/a)$ *ranges over all consecutive pairs of labels.*

We have discovered an equivalent definition of (lower) Christoffel words. (See Exercise 1.7 for details.)

Definition 1.4. Suppose $a \perp b$. Consider the Cayley graph of $\mathbb{Z}/(a+b)\mathbb{Z}$ with generator b. It is a cycle, with vertices $0, b, 2b, 3b, \ldots, a, 0 \bmod (a+b)$. Starting from zero and proceeding in the order listed above,

(i) label those edges (s,t) satisfying $s < t$ by x;
(ii) label those edges (s,t) satisfying $s > t$ by y;
(iii) read edge-labels in the prescribed order, i.e., $0 \xrightarrow{x} b \xrightarrow{*} \cdots \xrightarrow{*} a \xrightarrow{y} 0$.

The **lower Christoffel word** of slope b/a is the word $x \cdots y$ formed above.

Example. Pick $a = 7$ and $b = 4$. Figure 4 shows the Cayley graph of $\mathbb{Z}/11\mathbb{Z}$ with generator 4 and edges $u \to v$ labelled x or y according to whether or not $u < v$. Reading the edges clockwise from 0 yields the word $xxyxxyxxyxy$, which is the Christoffel word of slope 4/7 (see Figure 2).

Remark. Had we chosen the generator a instead of b for $\mathbb{Z}/(a+b)\mathbb{Z}$ and swapped the roles of x and y in Definition 1.4, the resulting word would have been the upper Christoffel word of slope b/a. (This fact is immediate after Proposition 4.2 but perhaps difficult to see before it.)

Exercise 1.6. Suppose $a \perp b$. Let (i,j) be the point on the Christoffel path path from $(0,0)$ to (a,b) with label t/a. Then $t \equiv (i+j)b \bmod (a+b)$ and $t \equiv \big((a-i)+(b-j)\big)a \bmod (a+b)$.

Exercise 1.7. Let w be the Christoffel word of slope b/a. Let w_k denote the $(k+1)$st letter of w.

(a) $w_k = x$ if and only if $kb \bmod (a+b) < (k+1)b \bmod (a+b)$.
(b) $w_k = y$ if and only if the set $\{kb+1, kb+2, \ldots, kb+b\}$ contains a multiple of $a+b$.

Exercise 1.8 ([Sim2004, Theorem 3]). Suppose $a \perp b$. Denote the Christoffel word of slope b/a by $w = w_0 w_1 \cdots w_{a+b-1}$, where $w_i \in \{x, y\}$ for $0 \leq i < a+b$. If $0 < c < a+b$ is such that $ca \equiv -1 \bmod (a+b)$, show that

$$\big\{jc \bmod (a+b) \mid j = 0, 1, \ldots, a-1\big\} = \big\{k \bmod (a+b) \mid w_k = x\big\}.$$

CHAPTER 2

Christoffel Morphisms

In this chapter we introduce the monoid of Christoffel morphisms and exhibit a minimal set of generators for the monoid. See also Chapter 2 of [Lot2002].

1. Christoffel morphisms

Definition 2.1. A **Christoffel morphism** is an endomorphism of the free monoid $\{x,y\}^*$ that sends each Christoffel word onto a conjugate of a Christoffel word.

Note that the set of Christoffel morphisms is closed under composition since any endomorphism of $\{x,y\}^*$ maps conjugate words to conjugate words (Exercise 2.1).

If G is an endomorphism of $\{x,y\}^*$ and $w = a_0a_1\cdots a_r$ is a word in $\{x,y\}^*$ with $a_0, a_1\ldots, a_r \in \{x,y\}$, then

$$G(w) = G(a_0a_1\cdots a_r) = G(a_1)G(a_2)\cdots G(a_r).$$

Therefore, G is determined by the images of x and y, so we identify G with the ordered pair $\big(G(x), G(y)\big)$.

Example. We use the above notation to define the following five important endomorphisms of $\{x,y\}^*$.

$$\begin{gathered}\mathbf{G} = (x, xy), \quad \mathbf{D} = (yx, y),\\ \widetilde{\mathbf{G}} = (x, yx), \quad \widetilde{\mathbf{D}} = (xy, y),\\ \mathbf{E} = (y, x).\end{gathered}$$

It is easy to see that these five morphisms are injective on $\{x,y\}^*$ (Exercise 2.4). The remainder of this section is devoted to showing that they are also Christoffel morphisms.

Lemma 2.2. *The morphism* $\mathbf{G}$ *maps the Christoffel word of slope* b/a *to the Christoffel word of slope* $b/(a+b)$. *The morphism* $\widetilde{\mathbf{D}}$ *maps the Christoffel word of slope* b/a *to the Christoffel word of slope* $(a+b)/a$.

PROOF. We first prove the result for $\mathbf{G}$. Suppose $a \perp b$. The Christoffel word w of slope b/a, by definition, encodes the steps of the Christoffel path from $(0,0)$ to (a,b): the letter x encodes the step $\vec{e}_1$ and the letter y encodes the step $\vec{e}_2$, where $\vec{e}_1$ and $\vec{e}_2$ are the standard basis vectors of $\mathbb{R}^2$. Since $\mathbf{G}$ maps x to x and y to xy, the word $\mathbf{G}(w)$ corresponds to the path obtained from the Christoffel path from $(0,0)$ to (a,b) by replacing each step $\vec{e}_2$ by the two steps $\vec{e}_1$ and $\vec{e}_2$. We will show that this path is the Christoffel path from $(0,0)$ to $(a+b,b)$, implying that $\mathbf{G}(w)$ is the Christoffel word of slope $b/(a+b)$.

Define a linear transformation $\mathcal{G}\colon \mathbb{R}^2 \to \mathbb{R}^2$ by $\mathcal{G}(c,d) = (c+d,d)$ for all $(c,d) \in \mathbb{R}^2$. Let W denote the Christoffel path from $(0,0)$ to (a,b). Then $\mathcal{G}(W)$ is a

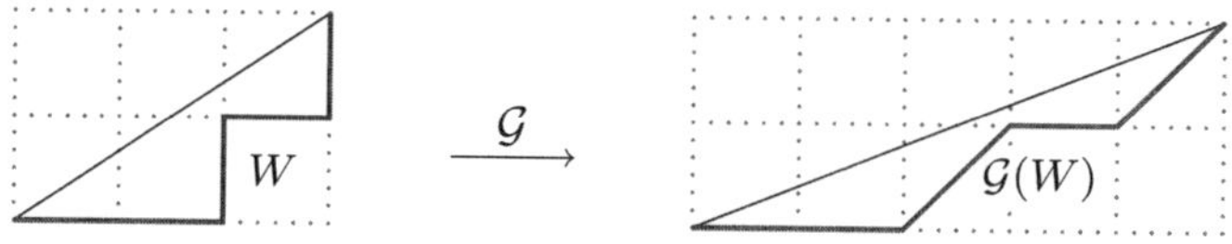

FIGURE 1. The image of a Christoffel path W under the linear transformation $\mathcal{G}$.

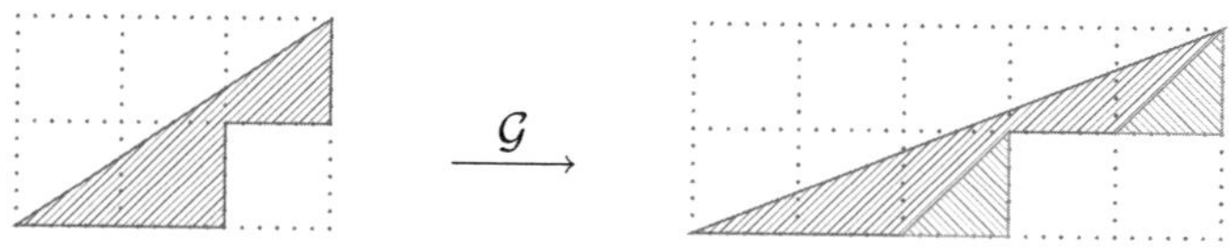

FIGURE 2. The image of the region between the line segment and the Christoffel path from $(0,0)$ to $(3,2)$ under the map $\mathcal{G}$.

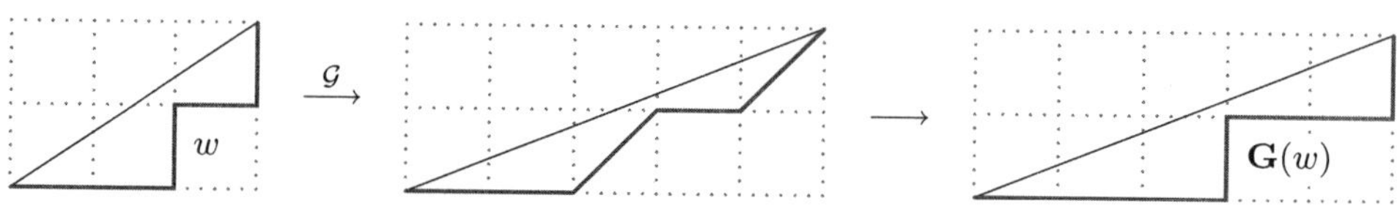

FIGURE 3. The geometric interpretation of the morphism $\mathbf{G}$.

path in the integer lattice $\mathbb{Z}\times\mathbb{Z}$ consisting of steps $\mathcal{G}(\vec{e}_1)=\vec{e}_1$ and $\mathcal{G}(\vec{e}_2)=\vec{e}_1+\vec{e}_2$. See Figure 1. We argue that the path obtained from $\mathcal{G}(W)$ by replacing the steps $\vec{e}_1+\vec{e}_2$ with the pair of steps $\vec{e}_1$ and $\vec{e}_2$ is the Christoffel path from $(0,0)$ to $(a+b,b)$.

Let R denote the region between the Christoffel path W and the line segment from $(0,0)$ to (a,b). Then there are no integer points in the interior of the region $\mathcal{G}(R)$ because $\mathcal{G}$ is a linear transformation and the region R contains no integer points in its interior. Therefore, there are no integer points in the interior of the region obtained from $\mathcal{G}(R)$ by adjoining the triangles with vertices $\vec{v}$, $\vec{v}+\vec{e}_1$ and $\vec{v}+(\vec{e}_1+\vec{e}_2)$ whenever $\vec{v}$ and $\vec{v}+(\vec{e}_1+\vec{e}_2)$ are in $\mathcal{G}(R)$. See Figure 2. The boundary of this new region consists of the line segment from $(0,0)$ to $(a+b,b)$ and the path P obtained from the path $\mathcal{G}(W)$ by replacing the steps $\vec{e}_1+\vec{e}_2$ with the steps $\vec{e}_1$ and $\vec{e}_2$. Also, $(a+b)\perp b$ since $a\perp b$ (see Exercise 1.2 or 1.4). Therefore, P is the Christoffel path from $(0,0)$ to $(a+b,b)$. Moreover, P is the path encoded by the word $\mathbf{G}(w)$ since P is obtained from the Christoffel path from $(0,0)$ to (a,b) by replacing each step $\vec{e}_2$ with the steps $\vec{e}_1$ and $\vec{e}_2$ (see Figure 3). Hence, $\mathbf{G}(w)$ is the Christoffel word of slope $b/(a+b)$.

The proof that $\widetilde{\mathbf{D}}(w)$ is a Christoffel word for any Christoffel word w is similar: define a linear transformation $\widetilde{\mathcal{D}}\colon\mathbb{R}^2\to\mathbb{R}^2$ by $\widetilde{\mathcal{D}}(c,d)=(c,c+d)$ for all $(c,d)\in\mathbb{R}^2$ and argue, as above, that $\widetilde{\mathbf{D}}$ maps the Christoffel word of slope b/a to the Christoffel word of slope $(a+b)/a$. □

Tracing backwards through the proof we also have the following result.

Corollary 2.3. *If u is a Christoffel word of slope at most one, then the unique word w such that $\mathbf{G}(w)=u$ is a Christoffel word. If u is a Christoffel word of slope at least one, then the unique word w such that $\widetilde{\mathbf{D}}(w)=u$ is a Christoffel word.*

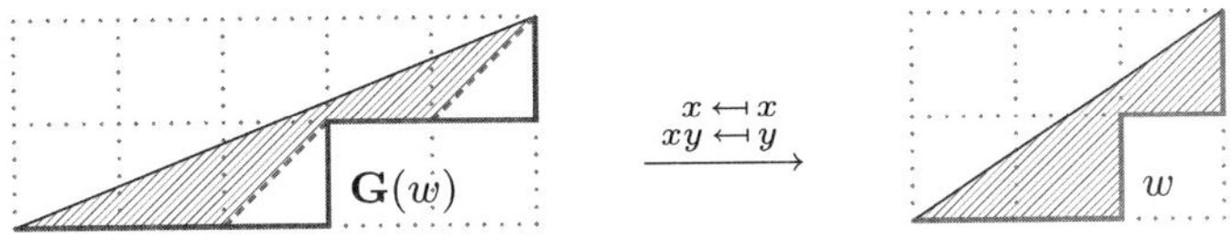

FIGURE 4. Christoffel words $\mathbf{G}(w)$ of slope less than 1 come from Christoffel words w.

The next lemma relates the image of $\mathbf{G}$ with that of $\widetilde{\mathbf{G}}$. We will use it again in future chapters.

Lemma 2.4. *For every word $w \in \{x,y\}^*$, there exists a word $u \in \{x,y\}^*$ such that $\mathbf{G}(w) = xu$ and $\widetilde{\mathbf{G}}(w) = ux$, and a word $v \in \{x,y\}^*$ such that $\mathbf{D}(w) = yv$ and $\widetilde{\mathbf{D}}(w) = vy$.*

PROOF. We prove the result for $\mathbf{G}$ and $\widetilde{\mathbf{G}}$; the proof for $\mathbf{D}$ and $\widetilde{\mathbf{D}}$ is the same. Proceed by induction on the length of w. If $|w| = 1$, then w is either x or y. So $\mathbf{G}(x) = x = \widetilde{\mathbf{G}}(x)$ with $u = \epsilon$, or $\mathbf{G}(y) = xy$ and $\widetilde{\mathbf{G}}(y) = yx$ with $u = y$. This establishes the base case of the induction. Let w be a word of length $r \geq 1$ and suppose the claim holds for all words of length less than r. If $w = xw'$ for some $w' \in \{x,y\}^*$, then the induction hypothesis implies there exists a word $u' \in \{x,y\}$ such that $\mathbf{G}(w') = xu'$ and $\widetilde{\mathbf{G}}(w') = u'x$. Therefore, $\mathbf{G}(w) = \mathbf{G}(x)\mathbf{G}(w') = xxu'$ and $\widetilde{\mathbf{G}}(w) = \widetilde{\mathbf{G}}(x)\widetilde{\mathbf{G}}(w') = xu'x$. Taking $u = xu'$, we are done. Suppose, instead, that $w = yw'$. Then the induction hypothesis implies there exists $u' \in \{x,y\}$ such that $\mathbf{G}(w') = xu'$ and $\widetilde{\mathbf{G}}(w') = u'x$. Here, $\mathbf{G}(w) = \mathbf{G}(y)\mathbf{G}(w') = xyxu'$ and $\widetilde{\mathbf{G}}(w) = \widetilde{\mathbf{G}}(y)\widetilde{\mathbf{G}}(w') = yxu'x$, so take $u = yxu'$. □

Corollary 2.5. *The morphisms $\mathbf{G}$, $\mathbf{D}$, $\widetilde{\mathbf{G}}$ and $\widetilde{\mathbf{D}}$ are Christoffel morphisms.*

PROOF. By Lemma 2.2, $\mathbf{G}$ and $\widetilde{\mathbf{D}}$ map Christoffel words to Christoffel words. Hence, they are Christoffel morphisms. We prove that $\widetilde{\mathbf{G}}$ is a Christoffel morphism; the same argument proves that $\mathbf{D}$ is a Christoffel morphism. Let w be a Christoffel word. Lemma 2.4 implies there exists a word $u \in \{x,y\}^*$ such that $\mathbf{G}(w) = xu$ and $\widetilde{\mathbf{G}}(w) = ux$. Therefore, $\mathbf{G}(w)$ and $\widetilde{\mathbf{G}}(w)$ are conjugate words. Since $\mathbf{G}(w)$ is a Christoffel word, $\widetilde{\mathbf{G}}(w)$ is a conjugate of a Christoffel word; that is, $\widetilde{\mathbf{G}}$ is a Christoffel morphism. □

We now turn to proving that $\mathbf{E}$ is a Christoffel morphism.

Lemma 2.6. *The morphism $\mathbf{E}$ maps lower Christoffel words of slope r onto upper Christoffel words of slope $1/r$.*

PROOF. This follows from an argument similar to that of Lemma 2.2, by using reflection about the line $x = y$. See Figure 5. □

Lemma 2.7 (Cohn [Coh1972], de Luca, Mignosi [dLM1994]). *Suppose $a \perp b$. The lower and upper Christoffel words of slope b/a are conjugates.*

PROOF. Suppose $a \perp b$ and let w be the Christoffel word of slope b/a. The word ww encodes a path in $\mathbb{Z} \times \mathbb{Z}$ from $(0,0)$ to $(2a, 2b)$. It consists of two copies of the Christoffel path of slope b/a, the first starting at the origin and the second starting at (a,b). See Figure 6.

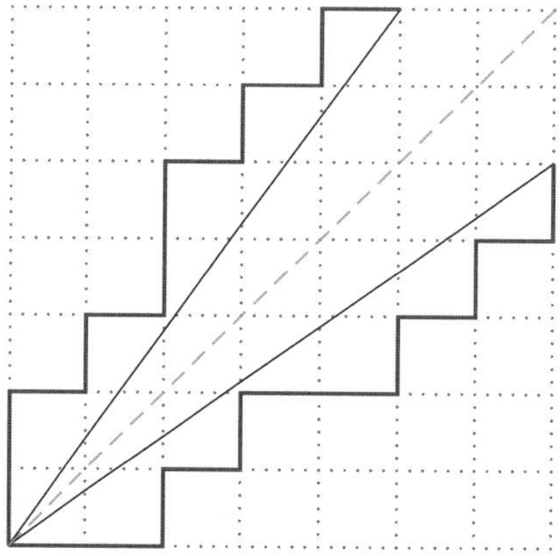

FIGURE 5. The geometric interpretation of the morphism **E** is reflection about the line $x = y$.

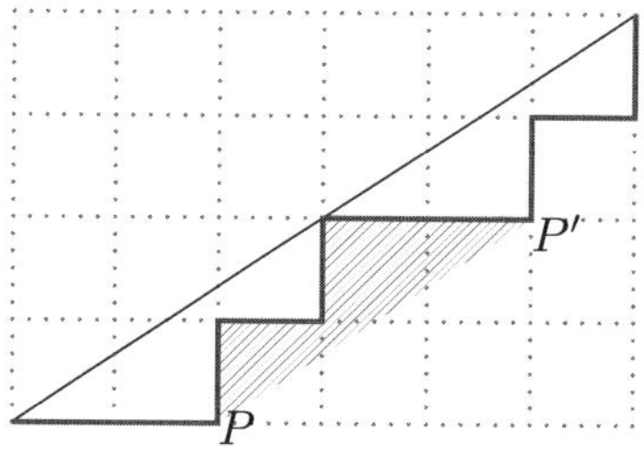

FIGURE 6. The path in $\mathbb{Z} \times \mathbb{Z}$ corresponding to the word $ww = xxyxyxxyxy$. The factor corresponding to the path from P to P' is $w' = yxyxx$. There are no integer points in the shaded region, so w' is an upper Christoffel word.

Let P denote the point on the first copy of the Christoffel path that is farthest (vertically) from the line segment defining the Christoffel path. By Lemma 1.3, this distance is $(a+b-1)/a$. Let P' denote the corresponding point on the translated copy of the Christoffel path. Then P and P' determine a word $w' \in \{x,y\}^*$ by encoding the part of the path from P to P' as a word in the letters x, y. Note that w' is a factor of ww of length equal to that of w. Since w' is a factor of ww and $l(w') = l(w)$, the words w' and w are conjugate (see Exercise 2.3). It remains to show that w' is the upper Christoffel word of slope b/a. We will argue that there is no other integer point in the region (shaded in Figure 6) enclosed by the line segment PP' and the path from P to P'. The argument is illustrated in Figure 7. Suppose $(i, j-1)$ is an integer point directly below a point (i,j) on the path from P to P', with $i > 0$. Then $(i-1, j)$ is also a point on the path. Suppose the point $(i, j-1)$ lies above or on the segment PP'. Then the vertical distance from $(i, j-1)$ to the segment from $(0,0)$ to $(2a, 2b)$ is at most the vertical distance from this segment to P, by the choice of P. The former is $ib/a - (j-1)$ and the latter is $(a+b-1)/a$. That is,

$$\frac{ib-(j-1)a}{a} \leq \frac{a+b-1}{a}.$$

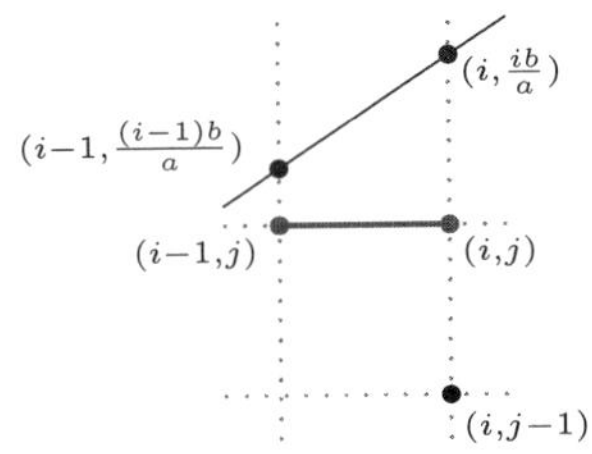

FIGURE 7. The points (i, j) and $(i-1, j)$ are on a Christoffel path, whereas $(i, j-1)$ is not.

Equivalently, $\big((i-1)b - ja\big)/a \le -1/a$. But $\big((i-1)b - ja\big)/a$ is nonnegative because it is the distance from the point $(i-1, j)$ to the segment from $(0,0)$ to $(2a, 2b)$. This is a contradiction, so there is no integer point within the region enclosed by the line segment PP' (of slope b/a) and the path from P to P'. That is, w' is the upper Christoffel word of slope b/a. □

Theorem 2.8. *The morphisms* $\mathbf{G}, \mathbf{D}, \widetilde{\mathbf{G}}, \widetilde{\mathbf{D}}, \mathbf{E}$ *are Christoffel morphisms.*

PROOF. The first four morphisms are Christoffel morphisms by Corollary 2.5. It remains to show that $\mathbf{E}$ is a Christoffel morphism. This follows from the previous two results: if w is a Christoffel word, then $\mathbf{E}(w)$ is an upper Christoffel word, which is a conjugate of the corresponding lower Christoffel word. □

Exercise 2.1. Prove that if f is an endomorphism of a free monoid A^* and w and w' are conjugate words in A^*, then $f(w)$ and $f(w')$ are conjugate.

Exercise 2.2. Suppose A is a finite set. Let $\langle A \rangle$ denote the free group generated by A and let A^* denote the free monoid generated by A. Prove that any $w, w' \in A^*$ are conjugate in $\langle A \rangle$ (in the group-theoretic sense) if and only if w, w' are conjugate in A^* (in the word-theoretic sense).

Exercise 2.3. Suppose w and w' are words of the same length. If w' is a factor of ww, then w and w' are conjugate.

Exercise 2.4. Prove that the morphisms $\mathbf{G}$, $\mathbf{D}$, $\widetilde{\mathbf{G}}$, $\widetilde{\mathbf{D}}$ are injective. (*Hint*: Study a minimal counterexample, a pair of words $u \neq v$ with total length $|u| + |v|$.)

Exercise 2.5. Complete the proof of Lemma 2.4: For every word $w \in \{x, y\}^*$, there exists a word $v \in \{x, y\}^*$ such that $\mathbf{D}(w) = yv$ and $\widetilde{\mathbf{D}}(w) = vy$.

Exercise 2.6. Recall that the reversal of a word $w = a_0 a_1 \cdots a_r$ is the word $\widetilde{w} = a_r \cdots a_1 a_0$. Let w be a word over $\{x, y\}^*$. Prove that

$$\widetilde{\mathbf{G}(w)} = \widetilde{\mathbf{G}}(\widetilde{w}) \quad \text{and} \quad \widetilde{\mathbf{D}(w)} = \widetilde{\mathbf{D}}(\widetilde{w}).$$

Exercise 2.7. If $u \in \{x, y\}^*$ is a palindrome, then $\mathbf{G}(u)x$ and $\mathbf{D}(u)y$ are palindromes.

Exercise 2.8 (A stronger version of Lemma 2.4). If $w \in \{x, y\}^*$ is a palindrome, then there exist palindromes $u, v \in \{x, y\}^*$ such that $\mathbf{G}(w) = xu$ and $\widetilde{\mathbf{G}}(w) = ux$, and $\mathbf{D}(w) = yv$ and $\widetilde{\mathbf{D}}(w) = vy$.

2. Generators

The following theorem gives a manageable characterization of the monoid of Christoffel morphisms. We will see in Section 2 that it implies a very close relationship between the set of Christoffel morphisms and the set of Christoffel words. References in the language of Sturmian words and Sturmian morphisms include [MS1993], [Lot2002, Chapter 2], [WW1994] and [BdLR2008].

Theorem 2.9. *The monoid of Christoffel morphisms is generated by* $\mathbf{G}$, $\mathbf{D}$, $\widetilde{\mathbf{G}}$, $\widetilde{\mathbf{D}}$ *and* $\mathbf{E}$.

In fact, $\mathbf{D} = \mathbf{E} \circ \mathbf{G} \circ \mathbf{E}$ and $\widetilde{\mathbf{G}} = \mathbf{E} \circ \widetilde{\mathbf{D}} \circ \mathbf{E}$, but it will simplify the proof to retain these superfluous generators. The proof makes frequent use of the following easy fact, so we separate it as a lemma.

Lemma 2.10. *If w is a Christoffel word or a conjugate of a Christoffel word, then xx and yy cannot both be factors of w.*

Proof. Indeed, let w be the Christoffel word of slope b/a. Figure 8 indicates a geometric proof of the following statements. If $b/a < 1$, then w begins with xx and yy is not a factor of w. In the case $b/a > 1$, w ends with yy and xx is not a factor of w. The only nontrivial Christoffel word with neither xx nor yy is $C(1,1) = xy$. Since w begins with an x and ends with a y, xx and yy are not both factors of the square w^2. Finally, since every conjugate of w appears as a factor of w^2, the property holds for conjugates of Christoffel words as well. □

Proof of Theorem 2.9. In five steps.

(1) *A Christoffel morphism f is* **nonerasing**, *that is, the length of $f(w)$ is at least the length of w.*

An erasing morphism f must necessarily send x or y to the empty word. In the first case, $f(xyy)$ is not primitive, in the second $f(xxy)$ is not primitive. On the other hand, all conjugates of a Christoffel word are primitive (Exercise 1.1).

(2) *If f is a nonidentity Christoffel morphism, then $f(x)$ and $f(y)$ must begin or end by the same letter.*

Assume $f(x)$ begins by x (study $\mathbf{E} \circ f$ otherwise) and $f(y)$ begins by y. There are two possibilities:

(i) Suppose $f(x)$ ends by y. Either $f(y)$ ends by x or we are done. In the remaining case, $f(xy) = f(x)f(y)$ becomes $x \cdots yy \cdots x$, so xx and yy are both factors of every conjugate of $f(xy)$, save perhaps for $f(y)f(x) = y \cdots xx \cdots y$. On the other hand, $f(xy)$ is a conjugate of a Christoffel word u that is not equal to $f(x)f(y)$ or $f(y)f(x)$ since u begins by x and ends by y. Lemma 2.10 yields a contradiction.

No yy if $b/a < 1$. No xx if $b/a > 1$.

Figure 8. Impossible configurations in Christoffel words.

(ii) Suppose instead $f(x)$ ends by x and $f(y)$ ends by y. Note that in this case, xx is a factor of $f(xxy) = f(x)f(x)f(y) = (x\cdots x)(x\cdots x)(y\cdots y)$. Hence, yy is not a factor of $f(xxy)$ by the lemma. In particular, yy is a factor of neither $f(x)$ nor $f(y)$. Similarly, by considering $f(xyy)$, we see that xx is a factor of neither $f(x)$ nor $f(y)$. This in turn forces $f(x) = (xy)^i x$, $f(y) = y(xy)^j$ and $f(xy) = (xy)^{i+j+1}$. Whence $i + j = 0$ (Exercise 1.1). Then f is the identity morphism, contrary to our hypothesis.

(3) *If f is a nonidentity Christoffel morphism, then there exists a morphism $g\colon \{x,y\}^* \to \{x,y\}^*$ and an $\mathbf{H} \in \{\mathbf{G}, \mathbf{D}, \widetilde{\mathbf{G}}, \widetilde{\mathbf{D}}\}$ such that $f = \mathbf{H} \circ g$.*

A nonempty word w on $\{x,y\}$ belongs to $\{x,xy\}^*$ (i.e., is a word on the "letters" x and xy) if and only if w begins by x and does not contain the factor yy. Similar descriptions hold for words in $\{y,xy\}^*$, $\{x,yx\}^*$ and $\{xy,y\}^*$. We argue that the image of f belongs to one of these monoids. Since $\mathbf{G}$, $\mathbf{D}$, $\widetilde{\mathbf{G}}$ and $\widetilde{\mathbf{D}}$ are injective morphisms (Exercise 2.4), with images in the respective monoids above, this will allow us to compose f with $\mathbf{G}^{-1}$, $\mathbf{D}^{-1}$, $\widetilde{\mathbf{G}}^{-1}$ or $\widetilde{\mathbf{D}}^{-1}$, respectively to find g.

Since $f(xy)$ is a conjugate of a Christoffel word, xx and yy are not both factors of $f(xy)$. Assuming yy is not a factor of $f(xy)$, it follows that yy is a factor of neither $f(x)$ nor $f(y)$. By Step 2, $f(x)$ and $f(y)$ must then begin or end by the same letter, setting up several cases to check.

(i) If $f(x)$ and $f(y)$ both begin by x, then the image of f is a subset of $\{x,xy\}^*$. Therefore, $\mathbf{G}^{-1} \circ f = g$ is also a morphism of $\{x,y\}^*$.

(ii) If $f(x)$ and $f(y)$ both begin by y, then neither may end by y (on account of the lemma and our assumption that yy is not a factor of $f(xy)$). Thus $f(x)$ and $f(y)$ both end by x and neither contain yy as a factor. That is, $f(x), f(y) \in \{x,yx\}^*$ and $\widetilde{\mathbf{G}}^{-1} \circ f$ is a morphism of $\{x,y\}^*$.

(iii) The cases where $f(x)$ and $f(y)$ end by the same letter are handled analogous to the cases above.

(4) *In the composition $f = \mathbf{H} \circ g$ built above, g is a Christoffel morphism.*

We now have that $f = \mathbf{H} \circ g$, with $\mathbf{H} \in \{\mathbf{G}, \mathbf{D}, \widetilde{\mathbf{G}}, \widetilde{\mathbf{D}}\}$, and that f sends Christoffel words onto conjugates of Christoffel words. We aim to show that g does as well. We analyze the case $\mathbf{H} = \mathbf{G}$, the rest being similar.

First, recall that if $\mathbf{G}(w)$ is a Christoffel word, then w is a Christoffel word too (Corollary 2.3). We must show that if $\mathbf{G}(w)$ is a conjugate of a Christoffel word then w is as well. This is now easy, for if $\mathbf{G}(w) = uv$ with vu a Christoffel word, then v begins by x and u ends by y. Moreover, by the definition of $\mathbf{G}$, u must begin by x and yy is a factor of neither u, v nor uv. This implies that $u, v \in \{x,xy\}^*$, so $\mathbf{G}^{-1}(u)$ and $\mathbf{G}^{-1}(v)$ are defined, $w = \mathbf{G}^{-1}(u)\,\mathbf{G}^{-1}(v)$ and $\mathbf{G}^{-1}(v)\,\mathbf{G}^{-1}(u)$ is a Christoffel word.

(5) *There exist $\mathbf{H}_i \in \{\mathbf{G}, \mathbf{D}, \widetilde{\mathbf{G}}, \widetilde{\mathbf{D}}\}$ such that $f = \mathbf{H}_1 \circ \cdots \circ \mathbf{H}_s$.*

From Step 4, $f = \mathbf{H}_1 \circ g$ for some Christoffel morphism g and some $\mathbf{H}_1 \in \{\mathbf{G}, \mathbf{D}, \widetilde{\mathbf{G}}, \widetilde{\mathbf{D}}\}$. Moreover, $|f(x)| + |f(y)| > |g(x)| + |g(y)|$. An induction on $|f(x)| + |f(y)|$ completes the proof. □

Remark. We have proved something a priori stronger.

Corollary 2.11 (Berthé, de Luca, Reutenauer [BdLR2008]**).** *A morphism f on $\{x,y\}^*$ is a Christoffel morphism if and only if $f(xy)$, $f(xxy)$ and $f(xyy)$ are conjugates of Christoffel words.*

CHAPTER 3

Standard Factorization

The first section of this chapter proves that every Christoffel word can be expressed as the product of two Christoffel words in a unique way, and the second section builds an infinite, complete binary tree whose vertices correspond to Christoffel words via this unique factorization.

1. The standard factorization

This section proves that every Christoffel word can be expressed as the product of two Christoffel words in a unique way. This factorization is called the standard factorization and was introduced by Jean-Pierre Borel and François Laubie [BL1993]. Most of the results in this section are due to them.

Given that $a \perp b$, recall the method of labelling the Christoffel path from $(0,0)$ to (a,b) for nontrivial Christoffel words. By Lemma 1.3, if a and b are nonzero, then there is a unique point C on this path having label $1/a$. We call C the **closest point** for the path. It is the lattice point on the Christoffel path from $(0,0)$ to (a,b) with minimum nonzero distance to the line segment from $(0,0)$ to (a,b).

Definition 3.1. Suppose $a \perp b$ with $a, b > 0$. The **standard factorization** of the Christoffel word w of slope b/a is the factorization $w = (w_1, w_2)$, where w_1 encodes the portion of the Christoffel path from $(0,0)$ to the closest point C and w_2 encodes the portion from C to (a,b).

Example. The standard factorization of the Christoffel word of slope $4/7$ is $(xxy, xxyxxyxy)$. See Figure 1.

Proposition 3.2. *If (w_1, w_2) is the standard factorization of a nontrivial Christoffel word, then w_1 and w_2 are Christoffel words.*

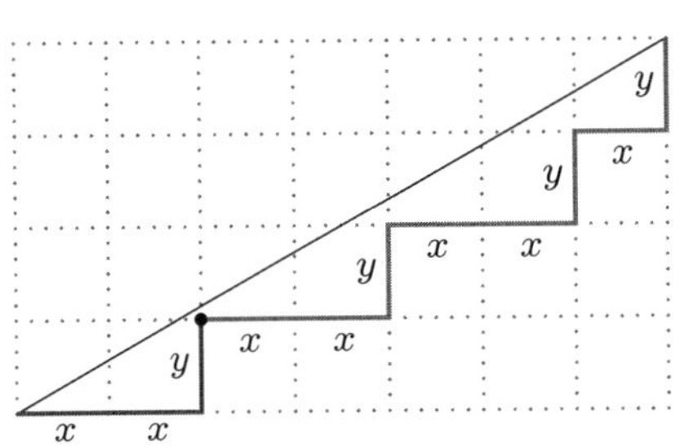

FIGURE 1. The closest point for the Christoffel path of slope $4/7$ occurs between the third and fourth steps, thus the standard factorization of $xxyxxyxxyxy$ is $(xxy, xxyxxyxy)$.

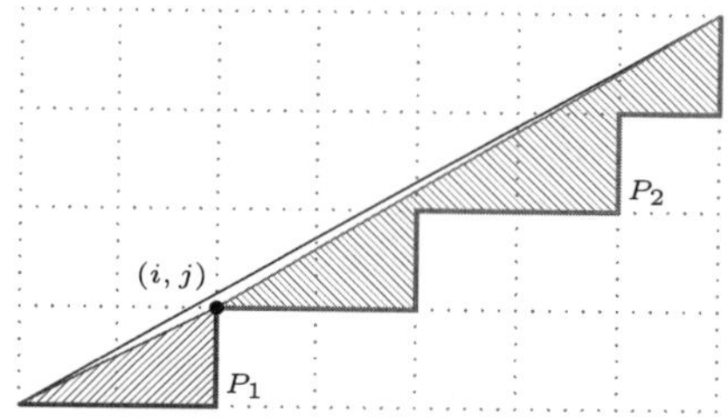

FIGURE 2. The standard factorization of a Christoffel word gives two Christoffel words.

PROOF. Suppose w is a Christoffel word of slope b/a and let (i, j) be the point on the Christoffel path from $(0, 0)$ to (a, b) labelled $1/a$. Then w_1 encodes the subpath P_1 from $(0, 0)$ to (i, j) and w_2 encodes the subpath P_2 from (i, j) to (a, b). See Figure 2.

Since (i, j) is the point on the Christoffel path that is closest to the line segment from $(0, 0)$ to (a, b) without being on the segment, no point of the Christoffel path besides $(0, 0)$, (a, b) and (i, j) lies in the triangle determined by these three points. See Figure 2. Let S_1 be the line segment from $(0, 0)$ to (i, j). Note that the region bounded by P_1 and S_1 contains no interior lattice points. Since, moreover, no integer points lie in the interior of the line segment S_1, it follows that $i \perp j$ (Exercise 1.2) and w_1 is the Christoffel word of slope j/i. Similarly, w_2 is the Christoffel word of slope $(b - j)/(a - i)$. $\square$

In fact, the standard factorization is the only factorization of a Christoffel word with this property.

Theorem 3.3 (Borel, Laubie [BL1993]). *A nontrivial Christoffel word w has a unique factorization $w = (w_1, w_2)$ with w_1 and w_2 Christoffel words.*

We present a proof suggested by Hugh Thomas.

PROOF. Let (w_1, w_2) denote the standard factorization of w. Recall that this factorization is obtained from cutting the Christoffel path at its closest point C. Suppose there is another factorization $w = (u, v)$ with u and v Christoffel words. See Figure 3. That is, $C' = (c, d)$ is another point on the path having no integer points in its corresponding regions (shaded in Figure 3) and satisfying $c \perp d$. We reach a contradiction by comparing triangles ABC and ABC' in Figure 3. Since

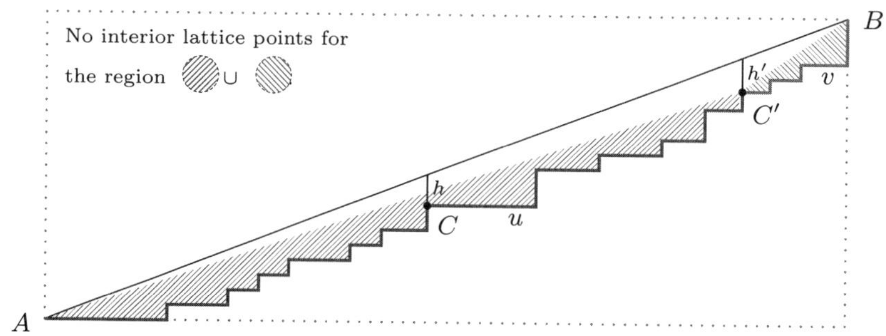

FIGURE 3. A Christoffel factorization $w = uv$ at cutpoint C'.

w_1, w_2 are Christoffel words, we know there are no integer lattice points in the interior of triangle ABC. Moreover, the only lattice points on its boundary occur at A, B and C. By *Pick's Theorem* (Exercise 3.1), we have

$$\text{area}\, ABC = i + \tfrac{1}{2}b - 1 = 0 + \tfrac{3}{2} - 1 = \tfrac{1}{2},$$

where i is the number of lattice points interior to ABC and b is the number of lattice points on its boundary. The same may be said for triangle ABC': since u, v are Christoffel words, the line segments AC' and BC' do not cross the Christoffel path for w; since w is a Christoffel word, this implies there are no interior lattice points in ABC'; there are only 3 boundary lattice points by the same reasoning. Now we have two triangles with the same base, the same area, but different heights. Contradiction. □

Finally, we record some additional facts about the factorization (w_1, w_2) that will be useful in what follows. Recall that $\mathrm{SL}_2(\mathbb{Z})$ is the group of invertible matrices with integer entries and determinant equal to 1.

Lemma 3.4. *Suppose (w_1, w_2) is the standard factorization of the Christoffel word w of slope b/a, where $a \perp b$. Then*

$$\begin{pmatrix} |w_1|_x & |w_2|_x \\ |w_1|_y & |w_2|_y \end{pmatrix} \in \mathrm{SL}_2(\mathbb{Z}).$$

PROOF. The point (i, j) of the Christoffel path labelled $1/a$ is $(|w_1|_x, |w_1|_y)$. Also, $(a-i, b-j) = (|w_2|_x, |w_2|_y)$. Since exactly three integer points lie in the triangle with vertices $(0,0)$ (i,j), (a,b), it follows from *Pick's Theorem* (see Exercises 3.1 and 3.2) that

$$\det \begin{pmatrix} i & j \\ a-i & b-j \end{pmatrix} = 1.$$

Therefore, the matrix is an element of $\mathrm{SL}_2(\mathbb{Z})$. □

Lemma 3.5. *Let w denote the Christoffel word of slope b/a and let (w_1, w_2) denote its standard factorization. Then $|w_1|b \equiv 1 \bmod (a+b)$ and $|w_2|a \equiv 1 \bmod (a+b)$. Moreover, $|w_1|$ and $|w_2|$ are relatively prime.*

PROOF. By Exercise 1.6, the point (i, j) on the Christoffel path from $(0,0)$ to (a, b) has label t/a, where t satisfies

$$t \equiv (i+j)b \bmod (a+b),$$
$$t \equiv \big((a-i)+(b-j)\big)a \bmod (a+b)$$

(recall that $|w|_x = a$ and $|w|_y = b$). Since $(|w_1|_x, |w_1|_y)$ is the closest point of the Christoffel path to the line segment from $(0,0)$ to (a,b), it has label $1/a$. Applying the above to $t = 1$ and the point $(i,j) = (|w_1|_x, |w_1|_y)$, we have $|w_1|b \equiv 1 \bmod (a+b)$ and $|w_2|a \equiv 1 \bmod (a+b)$.

It remains to show that $|w_1|$ and $|w_2|$ are relatively prime. By Corollary 3.4,

$$\begin{pmatrix} |w_1|_x & |w_2|_x \\ |w_1|_y & |w_2|_y \end{pmatrix} \in \mathrm{SL}_2(\mathbb{Z}).$$

This implies that

$$\det \begin{pmatrix} |w_1| & |w_2| \\ |w_1|_y & |w_2|_y \end{pmatrix} = \det \begin{pmatrix} |w_1|_x & |w_2|_x \\ |w_1|_y & |w_2|_y \end{pmatrix} = 1.$$

That is, there exist integers k and l such that $|w_1|k + |w_2|l = 1$, which implies $|w_1| \perp |w_2|$ (see *Bézout's Lemma* in Exercise 3.3). □

Exercise 3.1 (Pick's Theorem). Let P be a simple polygon (that is, the boundary of P has no self-intersections) with vertices in $\mathbb{Z} \times \mathbb{Z}$. Then the area of P is $i + \frac{1}{2}b - 1$, where i is the number of integer points in the interior of P and b is the number of integer points of the boundary of P. (*Hint*: Proceed by induction on the number of vertices of P.)

Exercise 3.2. Suppose i, j, k, l are positive integers. If no other integer points lie in the triangle with vertices $(0,0)$, (i,j), $(i+k, j+l)$, then

$$\det \begin{pmatrix} i & j \\ k & l \end{pmatrix} = 1.$$

(*Hint*: Use Pick's Theorem above and the fact that the determinant is twice the area of the triangle.)

Exercise 3.3 (Bézout's Lemma). Let a and b be positive integers. If the greatest common divisor of a and b is d, then there exist integers i and j such that $ia + jb = d$. Moreover, $a \perp b$ if and only if there exist integers i and j such that $ia + jb = 1$.

2. The Christoffel tree

We close this chapter with a description of the Christoffel tree, following [BL1993] and [BdL1997]. This is the infinite, complete binary tree whose root is labelled (x, y) and whose vertices are labelled by pairs (u, v) of words in $\{x, y\}^*$ subject to the following branching rules.

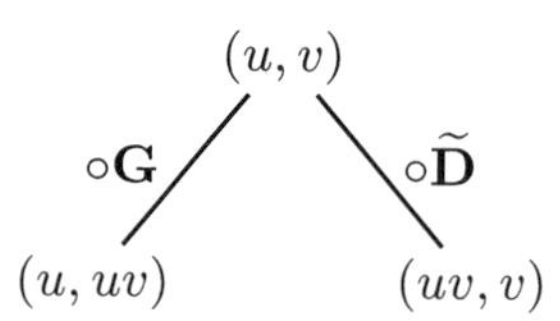

View the vertex (u, v) above as a morphism $(x, y) \overset{f}{\mapsto} (u, v)$. We have labelled the edges to indicate that $f = (u, v)$ has two branches, $f \circ \mathbf{G}$ and $f \circ \widetilde{\mathbf{D}}$. These rules were introduced by Gerard Rauzy in [Rau1984]. The first few levels of the Christoffel tree appear in Figure 4.

Theorem 3.6. *The Christoffel tree contains exactly once the standard factorization of each (lower) Christoffel word.*

Example. Recall that $(xxy, xxyxxyxy)$ is the standard factorization of the Christoffel word of slope $4/7$ (see Figure 1). It appears in Figure 4 at the fifth level as $(\mathbf{G} \circ \widetilde{\mathbf{D}} \circ \mathbf{G} \circ \mathbf{G})(x, y)$.

Proof of Theorem 3.6. In three steps.

(1) *Each vertex (u, v) on the tree has the property that u, v and uv are Christoffel words.*

We have seen that $\mathbf{G}$ and $\widetilde{\mathbf{D}}$ send Christoffel words to Christoffel words. Since each $f = (u, v)$ on the tree corresponds to a composition of $\mathbf{G}$s and $\widetilde{\mathbf{D}}$s, we get immediately that $u = f(x)$, $v = f(y)$ and $uv = f(xy)$ are Christoffel words.

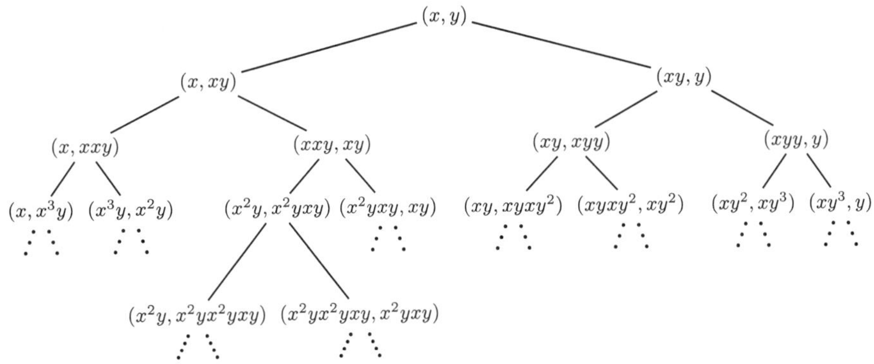

FIGURE 4. The Christoffel tree.

(2) *A vertex (u, v) on the Christoffel tree is the standard factorization of the Christoffel word uv.*

By Step 1, u, v and uv are Christoffel words. By Theorem 3.3, the only way to factor uv as Christoffel words is the standard factorization (u, v).

(3) *The standard factorization (w_1, w_2) of a Christoffel word w appears exactly once in the Christoffel tree.*

We demonstrate how to write

$$(w_1, w_2) = \big(\mathbf{H}_1 \circ \mathbf{H}_2 \circ \cdots \circ \mathbf{H}_r\big)(x, y)$$

for some $r \in \mathbb{N}$ and $\mathbf{H}_i \in \{\mathbf{G}, \widetilde{\mathbf{D}}\}$, thereby explicitly describing a path in the Christoffel tree from the root to the vertex (w_1, w_2). The argument is illustrated in the example following this proof. We need only argue that standard factorizations may be lifted via $\mathbf{G}$ or $\widetilde{\mathbf{D}}$. Specifically, we apply $\mathbf{G}^{-1}$ if w_1w_2 has slope less than 1 and $\widetilde{\mathbf{D}}^{-1}$ if w_1w_2 has slope greater than 1 (see Corollary 2.3). An example where the slope is less than 1 is illustrated in Figure 5.

The figure suggests that the closest point does not change under a change of basis. More precisely, we claim that if (u, v) is the standard factorization of the Christoffel word uv of slope less than 1, then $\big(\mathbf{G}^{-1}(u), \mathbf{G}^{-1}(v)\big)$ is the standard factorization of $\mathbf{G}^{-1}(uv)$. First, since yy is not a factor of uv (see the proof of Lemma 2.10), yy is not a factor of u or v. Hence, u and v are in the image of $\mathbf{G}$. Moreover, u and v are Christoffel words, so $\mathbf{G}^{-1}(u)$ and $\mathbf{G}^{-1}(v)$ are as well

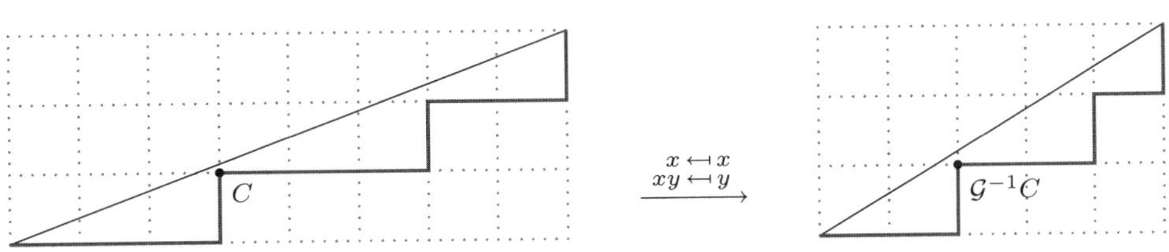

FIGURE 5. C is the closest point for the Christoffel path of $\mathbf{G}(xxyxxyxy) = xxxyxxxyxxy$ and $\mathcal{G}^{-1}(C)$ is the closest point for the Christoffel path of the word $xxyxxyxy$. (Here $\mathcal{G}$ is the linear transformation $(\vec{e}_1, \vec{e}_2) \mapsto (\vec{e}_1 + \vec{e}_2, \vec{e}_2)$ mimicing $\mathbf{G}$.)

(by Corollary 2.3). This is the standard factorization of $\mathbf{G}^{-1}(uv)$ by Theorem 3.3. The same argument works for $\widetilde{\mathbf{D}}$.

Finally, the fact that (w_1, w_2) can occur at most once in the Christoffel tree comes from the following property of binary trees. Each vertex describes a unique path back to the root, a finite sequence of statements of the form, "I was a left branch" or "I was a right branch." Since being a left branch corresponds to precomposition by $\mathbf{G}$ and being a right branch corresponds to precomposition by $\widetilde{\mathbf{D}}$, if (w_1, w_2) appears at two distinct vertices of the graph, then we have two expressions

$$(w_1, w_2) = \big(\mathbf{H}_1 \circ \mathbf{H}_2 \circ \cdots \circ \mathbf{H}_r\big)(x, y)$$
$$(w_1, w_2) = \big(\mathbf{H}'_1 \circ \mathbf{H}'_2 \circ \cdots \circ \mathbf{H}'_s\big)(x, y)$$

for some $r, s \in \mathbb{N}$ and $\mathbf{H}_i, \mathbf{H}'_i \in \{\mathbf{G}, \widetilde{\mathbf{D}}\}$. Since the only Christoffel word in the image of both $\mathbf{G}$ and $\widetilde{\mathbf{D}}$ is xy (corresponding to the root of the tree), it follows that $\mathbf{H}_1 = \mathbf{H}'_1$, $\mathbf{H}_2 = \mathbf{H}'_2$, …, $\mathbf{H}_r = \mathbf{H}'_s$. Therefore, both vertices of the graph describe the same (unique) path back to the root, contradicting the assumption that the two vertices are distinct. □

Example. We illustrate the ideas of Step 3 of the proof in Figure 6, marching upwards from the vertex $(xxxy, xxxyxxy)$ to the root (x, y).

Note that we have found a characterization of those Christoffel morphisms that preserve Christoffel words. Namely, $f\colon (x, y) \mapsto (w_1, w_2)$ is such a morphism if and only if (w_1, w_2) is a standard factorization of a Christoffel word.

Exercise 3.4. Let f be a Christoffel morphism. Prove that f takes Christoffel words to Christoffel words if and only if $f = (w_1, w_2)$, where (w_1, w_2) is the standard factorization of some Christoffel word. (*Hint*: One direction is Theorem 3.3. For the other direction, use the Christoffel tree to show that f is a composition of $\mathbf{G}$s and $\widetilde{\mathbf{D}}$s.)

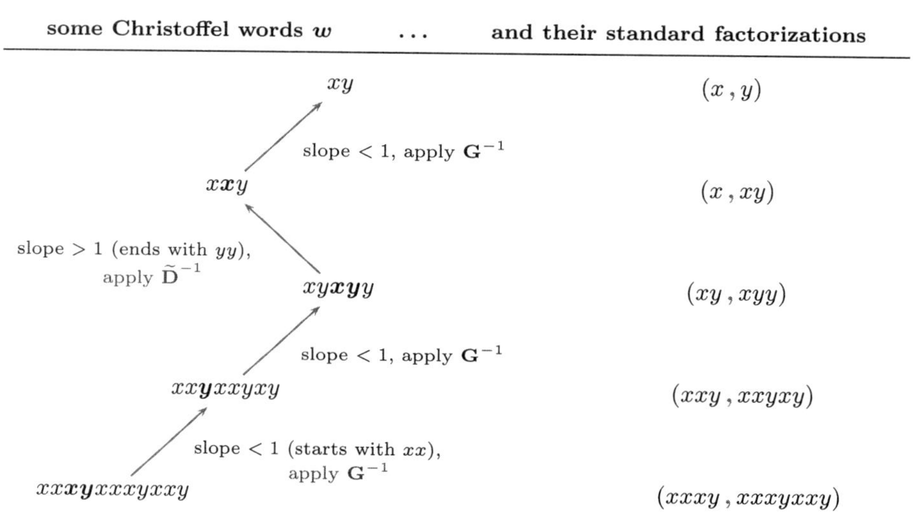

FIGURE 6. Paths in the Christoffel tree from (u, v) to the root (x, y) preserve the cutting points for standard factorizations.

CHAPTER 4

Palindromization

Recall that a word u is a **palindrome** if it is equal to its own reversal ($u = \tilde{u}$). This chapter begins with the observation that if w is a nontrivial Christoffel word, then $w = xuy$ with u a (possibly empty) palindrome. It continues by investigating the set of palindromes u for which xuy is a Christoffel word.

1. Christoffel words and palindromes

We prove that every nontrivial (lower) Christoffel word can be expressed as xuy with u a palindrome, and that the corresponding upper Christoffel word is yux.

Lemma 4.1. *Suppose $a \perp b$. Translation by the vector $\vec{e}_2 - \vec{e}_1$ and rotation about the point $(a/2, b/2)$ each map the interior points of the lower Christoffel path from $(0,0)$ to (a,b) onto the interior points of the upper Christoffel path from $(0,0)$ to (a,b).*

PROOF. *Translation*: Let Q be a point different from $(0,0)$ and (a,b) on the lower Christoffel path from $(0,0)$ to (a,b). Then the translated point $Q + (\vec{e}_2 - \vec{e}_1)$ is an integer point lying above the lower Christoffel path, and so it lies above the segment from $(0,0)$ to (a,b). Since there is no path in the integer lattice consisting of steps $\vec{e}_1$ and $\vec{e}_2$ that avoids Q and $Q+(\vec{e}_2-\vec{e}_1)$, and that has Q and $Q+(\vec{e}_2-\vec{e}_1)$ on opposite sides of the path, it follows that $Q + (\vec{e}_2 - \vec{e}_1)$ lies on the upper Christoffel path from $(0,0)$ to (a,b).

Rotation: Since there are no lattice points enclosed by the (upper or lower) Christoffel path and the segment from $(0,0)$ to (a,b), a half-turn about the midpoint of the line segment from $(0,0)$ to (a,b) maps the lower Christoffel path to the upper Christoffel path. □

The following result is the consequence of the above geometric lemma. (Recall that a Christoffel word is nontrivial if its length is at least two.)

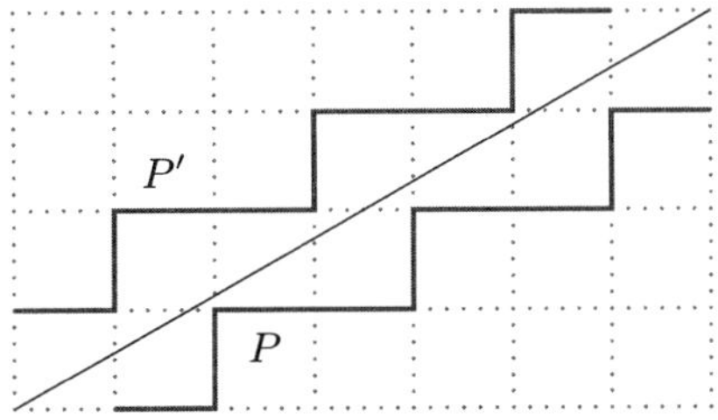

FIGURE 1. Translation by $\vec{e}_2 - \vec{e}_1$ maps P onto P'; rotation about $(7/2, 2)$ maps P onto the reverse of P'.

Proposition 4.2. *Suppose* $a \perp b$. *If* w *is a nontrivial lower Christoffel word of slope* b/a, *then* $w = xuy$ *with* u *a palindrome. If* w' *is the upper Christoffel word of slope* b/a, *then* $w' = yux$. *In particular,* $w' = \widetilde{w}$.

PROOF. Let w and w' be the nontrivial lower and upper Christoffel words of slope b/a, respectively. By construction any lower Christoffel word begins by x and ends by y, so $w = xuy$ for some $u \in \{x,y\}^*$. Similarly, $w' = yu'x$ for some $u' \in \{x,y\}^*$. The words u and u' correspond to the subpaths P and P' obtained from the lower and upper Christoffel paths from $(0,0)$ to (a,b), respectively, by removing the endpoints. By Lemma 4.1, P is a translate of P', so $u = u'$. Also by Lemma 4.1, a half-turn rotation maps P' onto P. Since rotation reverses the direction of P', it follows that $u = \tilde{u}' = \tilde{u}$. So u is a palindrome. Finally, $w' = yu'x = yux = y\tilde{u}x = \widetilde{w}$. □

2. Palindromic closures

We next determine those palindromes u for which xuy is a Christoffel word, following the work of Aldo de Luca [dL1997] and others. A function Pal that maps words to palindromes is defined and it will be shown that $x\,\mathrm{Pal}(v)y$ is a Christoffel word for every $v \in \{x,y\}^*$. It will be useful to have the terminology **palindromic prefix** and **palindromic suffix**, that is, a prefix (respectively, suffix) u of a word w such that u is a palindrome.

Proposition 4.3 (de Luca [dL1997]). *Let* w *be a word. Write* $w = uv$, *where* v *is the longest suffix of* w *that is a palindrome. Then* $w^+ = w\tilde{u}$ *is the unique shortest palindrome having* w *as a prefix.*

PROOF. The proof is left as an exercise (Exercise 4.5). □

Definition 4.4. Let w be a word. The word w^+ constructed in Proposition 4.3 is called the **(right) palindromic closure** of w.

Example. Let $w = yxyxxy$. The longest palindromic suffix of w is $v = yxxy$. Putting $u = yx$, we have $w^+ = w\tilde{u} = yxyxxyxy$ and w^+ is indeed a palindrome.

Definition 4.5 (de Luca [dL1997]). Define a function $\mathrm{Pal}\colon \{x,y\}^* \to \{x,y\}^*$ recursively as follows. For the empty word ϵ, let $\mathrm{Pal}(\epsilon) = \epsilon$. If $w = vz \in \{x,y\}^*$ for some $z \in \{x,y\}$, then let

$$\mathrm{Pal}(w) = \mathrm{Pal}(vz) = (\mathrm{Pal}(v)z)^+.$$

The word $\mathrm{Pal}(w)$ is called the **iterated palindromic closure** of w.

Example. We compute $\mathrm{Pal}(xyxx)$.

$$\mathrm{Pal}(x) = (\mathrm{Pal}(\epsilon)x)^+ = x^+ = x.$$

$$\mathrm{Pal}(xy) = (\mathrm{Pal}(x)y)^+ = (xy)^+ = xyx.$$

$$\mathrm{Pal}(xyx) = (\mathrm{Pal}(xy)x)^+ = \big((xyx)x\big)^+ = xyxxyx.$$

$$\mathrm{Pal}(xyxx) = (\mathrm{Pal}(xyx)x)^+ = \big((xyxxyx)x\big)^+ = xyxxyxxyx.$$

Note that the Christoffel word of slope $4/7$ is $xxyxxyxxyxy = x\,\mathrm{Pal}(xyxx)y$.

The map $w \mapsto \mathrm{Pal}(w)$ is injective. A complete proof is outlined in the exercises (Exercise 4.9). Briefly, the inverse map is obtained by taking the first letter after each palindromic prefix of $\mathrm{Pal}(w)$ (excluding $\mathrm{Pal}(w)$, but including the empty prefix ϵ). The fact that this procedure works follows from the observation that the only palindromic prefixes of $\mathrm{Pal}(w)$ are those obtained during the iterated construction of $\mathrm{Pal}(w)$.

Example. Suppose $\mathrm{Pal}(w) = xyxxyxxyx$. The palindromic prefixes of $\mathrm{Pal}(w)$ excluding $\mathrm{Pal}(w)$ are: ϵ; x; xyx; and $xyxxyx$. The first letter after these prefixes are: x; y; x; x. Therefore, $w = xyxx$. This agrees with the computation of $\mathrm{Pal}(xyxx)$ in the previous example. Moreover, from that computation we note that the words $\mathrm{Pal}(\epsilon)$, $\mathrm{Pal}(x)$, $\mathrm{Pal}(xy)$, $\mathrm{Pal}(xyx)$ and $\mathrm{Pal}(xyxx)$ are palindromic prefixes of $\mathrm{Pal}(xyxx)$, and that they are the only palindromic prefixes of $\mathrm{Pal}(xyxx)$. See Exercise 4.8.

The remainder of this section is devoted to proving a result that implies, among other things, that xuy is a Christoffel word if u is in the image of Pal. Before stating the full result we recall the definition of a period of a word. A positive integer p is a period of w if $w_i = w_{i+p}$ for all $1 \leq i \leq |w| - p$, where w_i denotes the ith letter of the word w. (Again, we allow $p \geq |w|$.)

Theorem 4.6 (Borel, Laubie [BL1993], de Luca [dL1997], Berthé, de Luca, Reutenauer [BdLR2008]). *Let $v \in \{x, y\}^*$. Then $w = x\,\mathrm{Pal}(v)y$ is a Christoffel word. If (w_1, w_2) is the standard factorization of w, then*

$$\mu(v) = \begin{pmatrix} |w_1|_x & |w_2|_x \\ |w_1|_y & |w_2|_y \end{pmatrix} \in \mathrm{SL}_2(\mathbb{Z}),$$

where $\mu\colon \{x, y\}^* \to \mathrm{SL}_2(\mathbb{Z})$ is the multiplicative monoid morphism defined by

$$\mu(x) = \begin{pmatrix} 1 & 1 \\ 0 & 1 \end{pmatrix} \quad \text{and} \quad \mu(y) = \begin{pmatrix} 1 & 0 \\ 1 & 1 \end{pmatrix},$$

and $\mathrm{Pal}(v)$ has relatively prime periods $|w_1|$ and $|w_2|$.

Remark. We provide a converse to this result in Proposition 4.14, namely, if w is a Christoffel word then $w = x\,\mathrm{Pal}(v)y$ for some $v \in \{x, y\}^*$.

Example. Let $w = xxyxxyxxyxy$ denote the Christoffel word of slope $4/7$. Note that $xyxxyxxyx$ has periods 3 and 8. In previous examples we have seen that $w = x\,\mathrm{Pal}(xyxx)y$ and that the standard factorization of w is $(w_1, w_2) = (xxy, xxyxxyxy)$. Therefore,

$$\mu(xyxx) = \begin{pmatrix} 1 & 1 \\ 0 & 1 \end{pmatrix}\begin{pmatrix} 1 & 0 \\ 1 & 1 \end{pmatrix}\begin{pmatrix} 1 & 1 \\ 0 & 1 \end{pmatrix}\begin{pmatrix} 1 & 1 \\ 0 & 1 \end{pmatrix} = \begin{pmatrix} 2 & 5 \\ 1 & 3 \end{pmatrix} = \begin{pmatrix} |w_1|_x & |w_2|_x \\ |w_1|_y & |w_2|_y \end{pmatrix}.$$

The proof is divided into three propositions. We begin by proving that $x\,\mathrm{Pal}(v)y$ is a Christoffel word.

The following formulae of Jacques Justin give a very useful method for computing $\mathrm{Pal}(v)$.

Lemma 4.7 (Justin [Jus2005]). *For any word $w \in \{x, y\}^*$,*

$$\begin{aligned} \mathrm{Pal}(xw) &= \mathbf{G}(\mathrm{Pal}(w)x) = \mathbf{G}\big(\mathrm{Pal}(w)\big)x, \\ \mathrm{Pal}(yw) &= \mathbf{D}(\mathrm{Pal}(w)y) = \mathbf{D}\big(\mathrm{Pal}(w)\big)y. \end{aligned} \tag{4.8}$$

A proof of this result is outlined in Exercise 4.11.

Example. We compute $\mathrm{Pal}(xyxx)$ using the formulae of Justin.

$$\begin{aligned}\mathrm{Pal}(xyxx) &= \mathbf{G}\big(\mathrm{Pal}(yxx)\big)x \\ &= \mathbf{G}\big(\mathbf{D}\big(\mathrm{Pal}(xx)\big)y\big)x \\ &= \mathbf{G}\big(\mathbf{D}(xx)y\big)x \\ &= \mathbf{G}(yxyxy)x \\ &= xyxxyxxyx.\end{aligned}$$

Proposition 4.9. *Suppose $v \in \{x, y\}^*$. Then $w = x\,\mathrm{Pal}(v)y$ is a Christoffel word.*

PROOF. Proceed by induction on $|v|$. Suppose the length of v is zero. Then $\mathrm{Pal}(v) = \epsilon$ and $w = xy$, which is a Christoffel word. Suppose that $x\,\mathrm{Pal}(v)y$ is a Christoffel word for all words v of length at most r and let $v' \in \{x, y\}^*$ be a word of length $r + 1$. If v' begins with x, then write $v' = xv$ for some $v \in \{x, y\}^*$. Then, by the formulae of Justin,

$$x\,\mathrm{Pal}(v')y = x\,\mathrm{Pal}(xv)y = x\mathbf{G}(\mathrm{Pal}(v)x)y = \mathbf{G}(x\,\mathrm{Pal}(v)y).$$

This is a Christoffel word because $x\,\mathrm{Pal}(v)y$ is a Christoffel word (by the induction hypothesis) and because $\mathbf{G}$ maps Christoffel words to Christoffel words (Lemma 2.2).

If $v' = yv$, then

$$x\,\mathrm{Pal}(v')y = x\,\mathrm{Pal}(yv)y = x\mathbf{D}(\mathrm{Pal}(v)y)y.$$

Lemma 2.4 implies there exists a word u such that $\widetilde{\mathbf{D}}(\mathrm{Pal}(v)y) = uy$ and $\mathbf{D}(\mathrm{Pal}(v)y) = yu$. The first equality together with $\widetilde{\mathbf{D}}(y) = y$ implies that $u = \widetilde{\mathbf{D}}\big(\mathrm{Pal}(v)\big)$. Therefore, $\mathbf{D}(\mathrm{Pal}(v)y) = y\widetilde{\mathbf{D}}\big(\mathrm{Pal}(v)\big)$. Hence,

$$x\,\mathrm{Pal}(v')y = x\mathbf{D}(\mathrm{Pal}(v)y)y = xy\widetilde{\mathbf{D}}\big(\mathrm{Pal}(v)\big)y = \widetilde{\mathbf{D}}(x\,\mathrm{Pal}(v)y).$$

This is a Christoffel word because $x\,\mathrm{Pal}(v)y$ is a Christoffel word (by the induction hypothesis) and because $\widetilde{\mathbf{D}}$ maps Christoffel words to Christoffel words (Lemma 2.2). □

We next prove that the entries of the matrix $\mu(v)$ are given by the numbers of occurrences of the letters x and y in the words w_1 and w_2, where (w_1, w_2) is the standard factorization of the Christoffel word $x\,\mathrm{Pal}(v)y$.

Proposition 4.10. *Suppose $v \in \{x, y\}^*$. If (w_1, w_2) is the standard factorization of the Christoffel word $x\,\mathrm{Pal}(v)y$, then*

$$\mu(v) = \begin{pmatrix} |w_1|_x & |w_2|_x \\ |w_1|_y & |w_2|_y \end{pmatrix}.$$

PROOF. We proceed by induction on the length of v. If $|v| = 0$, then $v = \epsilon$. So the Christoffel word $x\,\mathrm{Pal}(\epsilon)y$ is xy and its standard factorization is (x, y). Therefore,

$$\mu(\epsilon) = \begin{pmatrix} 1 & 0 \\ 0 & 1 \end{pmatrix} = \begin{pmatrix} |x|_x & |y|_x \\ |x|_y & |y|_y \end{pmatrix}.$$

This establishes the base case of the induction. Suppose the result holds for all words v of length at most $r-1 \geq 0$ and let v' be a word of length r. If v' begins with x, then $v' = xv$ for some $v \in \{x,y\}^*$. By the induction hypothesis,

$$\mu(v') = \mu(x)\mu(v) = \begin{pmatrix} 1 & 1 \\ 0 & 1 \end{pmatrix} \begin{pmatrix} |w_1|_x & |w_2|_x \\ |w_1|_y & |w_2|_y \end{pmatrix} = \begin{pmatrix} |w_1| & |w_2| \\ |w_1|_y & |w_2|_y \end{pmatrix},$$

where (w_1, w_2) is the standard factorization of $x \operatorname{Pal}(v)y$. Writing (w_1', w_2') for the standard factorization of the Christoffel word $x \operatorname{Pal}(v')y$, we would like to show that

$$\begin{pmatrix} |w_1| & |w_2| \\ |w_1|_y & |w_2|_y \end{pmatrix} = \begin{pmatrix} |w_1'|_x & |w_2'|_x \\ |w_1'|_y & |w_2'|_y \end{pmatrix}.$$

In view of Lemma 3.4 and Exercise 4.3, it suffices to show that $|w_1'|_x + |w_2'|_x = |w_1| + |w_2|$ and $|w_1'|_y + |w_2'|_y = |w_1|_y + |w_2|_y$. Equivalently, we need to show that $|x \operatorname{Pal}(v')y|_x = |x \operatorname{Pal}(v)y|$ and $|x \operatorname{Pal}(v')y|_y = |x \operatorname{Pal}(v)y|_y$. By the formulae of Justin (4.8),

$$x \operatorname{Pal}(v')y = x \operatorname{Pal}(xv)y = x\mathbf{G}\big(\operatorname{Pal}(v)\big)xy = \mathbf{G}(x \operatorname{Pal}(v)y).$$

Since $\mathbf{G} = (x, xy)$ replaces each letter of a word $m \in \{x,y\}^*$ with a word having exactly one occurrence of x, the number of occurrences of the letter x in $\mathbf{G}(m)$ is the length of m. Therefore,

$$|x \operatorname{Pal}(v')y|_x = |\mathbf{G}(x \operatorname{Pal}(v)y)|_x = |x \operatorname{Pal}(v)y|.$$

Since $\mathbf{G} = (x, xy)$ fixes the letter x and replaces y with a word having exactly one occurrence of y, we have $|\mathbf{G}(m)|_y = |m|_y$ for any word $m \in \{x,y\}^*$. Therefore,

$$|x \operatorname{Pal}(v')y|_y = |\mathbf{G}(x \operatorname{Pal}(v)y)|_y = |x \operatorname{Pal}(v)y|_y.$$

This completes the induction for words beginning with the letter x.

If, instead, v' begins with the letter y, then $v' = yv$ for some $v \in \{x,y\}^*$, and

$$\mu(v') = \mu(y)\mu(v) = \begin{pmatrix} 1 & 0 \\ 1 & 1 \end{pmatrix} \begin{pmatrix} |w_1|_x & |w_2|_x \\ |w_1|_y & |w_2|_y \end{pmatrix} = \begin{pmatrix} |w_1|_x & |w_2|_x \\ |w_1| & |w_2| \end{pmatrix},$$

where (w_1, w_2) is the standard factorization of $x \operatorname{Pal}(v)y$. As above, we need only show that $|x \operatorname{Pal}(v')y|_y = |x \operatorname{Pal}(v)y|$ and $|x \operatorname{Pal}(v')y|_x = |x \operatorname{Pal}(v)y|_x$. By the formulae of Justin (4.8), $x \operatorname{Pal}(v')y = x \operatorname{Pal}(yv)y = x\mathbf{D}\big(\operatorname{Pal}(v)\big)yy$. Since $\mathbf{D} = (yx, y)$, it follows that $|\mathbf{D}(m)|_y = |m|$ and $|\mathbf{D}(m)|_x = |m|_x$ for any word $m \in \{x,y\}^*$, so

$$|x \operatorname{Pal}(v')y|_y = \big|x\mathbf{D}\big(\operatorname{Pal}(v)\big)yy\big|_y = |\operatorname{Pal}(v)yy| = |x \operatorname{Pal}(v)y|,$$

$$|x \operatorname{Pal}(v')y|_x = \big|x\mathbf{D}\big(\operatorname{Pal}(v)\big)yy\big|_x = |x \operatorname{Pal}(v)|_x = |x \operatorname{Pal}(v)y|_x.$$

This completes the induction. □

We now turn to the computation of periods of the word $\operatorname{Pal}(v)$. The treatment here is based on the paper [BR2006] of Borel and Reutenauer. The following result determines a period of palindromes having palindromic prefixes.

Lemma 4.11 (de Luca [dL1997]**).** *If a palindrome u has a palindromic prefix $p \neq u$, then u has a period $|u| - |p|$.*

PROOF. Write $u = pv$ for some word v. Then $u = \tilde{v}p$ because u and p are palindromes. Since $u = pv$, we have $u_i = p_i$ for $0 \leq i < |p|$. And since $u = \tilde{v}p$, we have $u_{i+|v|} = p_i$ for $0 \leq i < |p|$. Therefore, $u_i = u_{i+|v|}$ for $0 \leq i < |p|$. That is, u has period $|v| = |u| - |p|$. □

Proposition 4.12. *Suppose $v \in \{x,y\}^*$. The word* $\operatorname{Pal}(v)$ *has periods $|w_1|$ and $|w_2|$, where (w_1, w_2) is the standard factorization of the Christoffel word $x\operatorname{Pal}(v)y$. Moreover, the periods $|w_1|$ and $|w_2|$ are relatively prime and $|\operatorname{Pal}(v)| = |w_1| + |w_2| - 2$.*

PROOF. Let (w_1, w_2) denote the standard factorization of $x\operatorname{Pal}(v)y$. Then w_1 and w_2 are Christoffel words by Proposition 3.2. There are two cases to consider.

Case 1: w_1 or w_2 is a trivial Christoffel word. If $w_1 = x$, then

$$\mu(v) = \begin{pmatrix} 1 & |w_2|_x \\ 0 & |w_2|_y \end{pmatrix}.$$

Since $\det(\mu(v)) = 1$, it follows that $|w_2|_y = 1$. So $w_2 = x^e y$ for some $e \in \mathbb{N}$. Hence, $\operatorname{Pal}(v) = x^e$, which has periods $|w_1| = 1$ and $|w_2| = e+1$, and length $|w_1| + |w_2| = 1 + (e+1) - 2 = e$. The same argument holds if $w_2 = y$.

Case 2: w_1 and w_2 are nontrivial Christoffel words. By Proposition 4.2, there exist palindromes u_1 and u_2 such that $w_1 = xu_1y$ and $w_2 = xu_2y$. Therefore, $\operatorname{Pal}(v) = u_1xyu_2$. The previous lemma implies that $\operatorname{Pal}(v)$ has periods $|\operatorname{Pal}(v)| - |u_1| = |u_2| + 2 = |w_2|$ and $|\operatorname{Pal}(v)| - |u_2| = |u_1| + 2 = |w_1|$. The fact that $|w_1|$ and $|w_2|$ are relatively prime follows from Lemma 3.5. □

Exercise 4.1. Given any endomorphism f of the monoid $\{x,y\}^*$ and any word $w \in \{x,y\}^*$, one has

$$\begin{pmatrix} |f(w)|_x \\ |f(w)|_y \end{pmatrix} = \begin{pmatrix} |f(x)|_x & |f(y)|_x \\ |f(x)|_y & |f(y)|_y \end{pmatrix} \begin{pmatrix} |w|_x \\ |w|_y \end{pmatrix}.$$

Exercise 4.2. Show that the monoid $\mathrm{SL}_2(\mathbb{Z}) \cap \mathbb{N}^{2\times 2}$ is generated by

$$\begin{pmatrix} 1 & 1 \\ 0 & 1 \end{pmatrix} \quad \text{and} \quad \begin{pmatrix} 1 & 0 \\ 1 & 1 \end{pmatrix},$$

which are the images of x and y under the morphism μ of Theorem 4.6.

Exercise 4.3 (**[Ran1973, BdL1997]**)**.** Two matrices

$$M = \begin{pmatrix} a & b \\ c & d \end{pmatrix} \text{ and } M' = \begin{pmatrix} a' & b' \\ c' & d' \end{pmatrix} \text{ in } \mathbb{N}^{2\times 2} \cap \mathrm{SL}_2(\mathbb{Z})$$

satisfy $a + b = a' + b'$ and $c + d = c' + d'$ if and only if $M = M'$.

Exercise 4.4. If $w \in \{x,y\}^*$, then $\mathbf{E}(\operatorname{Pal}(w)) = \operatorname{Pal}(\mathbf{E}(w))$. (*Hint*: First establish that $(\mathbf{E}(w))^+ = \mathbf{E}(w^+)$.)

Exercise 4.5. Prove Proposition 4.3. (*Hint*: Show that $w^+ = wv^{-1}\widetilde{w}$, where the product on the left is evaluated in the free group generated by x and y.)

Exercise 4.6. If $p \neq w^+$ is a palindromic prefix of w^+, then p is a (palindromic) prefix of w.

Exercise 4.7. Given two letters x and y and an integer $s > 0$, prove that:

(a) $\operatorname{Pal}(xy^s) = (xy)^s x$;
(b) $\operatorname{Pal}(x^s y) = \operatorname{Pal}(x^s) y \operatorname{Pal}(x^s) = x^s y x^s$;
(c) $\operatorname{Pal}(xy^s x) = \operatorname{Pal}(xy^s)\operatorname{Pal}(xy^s)$.

Exercise 4.8. Let w be a word and $z \in \{x,y\}$.

(a) If u is a prefix of w, then $\operatorname{Pal}(u)$ is a prefix of $\operatorname{Pal}(w)$.

(b) uz is a prefix of w if and only if $\operatorname{Pal}(u)z$ is a prefix of $\operatorname{Pal}(w)$.
(c) If p is a palindromic prefix of $\operatorname{Pal}(w)$, then $p = \operatorname{Pal}(u)$ for some prefix u of w.

(*Hint*: Proceed by induction on the length of w and use Exercise 4.6.)

Exercise 4.9 (Pal is injective). If $\epsilon = p_1, p_2, \ldots, p_r$ denote the sequence of palindromic prefixes of $\operatorname{Pal}(w)$ different than $\operatorname{Pal}(w)$ listed in order of increasing length, and if $z_1, z_2, \ldots, z_r \in \{x, y\}$ denote the letters in $\operatorname{Pal}(w)$ immediately following the prefixes $p_1, p_2, \ldots, p_r$ in $\operatorname{Pal}(w)$, then $w = z_1 z_2 \cdots z_r$. (*Hint*: Use Exercise 4.8.)

Exercise 4.10. If $w = vzu$, where u does not have an occurrence of the letter z, then

$$\operatorname{Pal}(wz) = \operatorname{Pal}(w)\operatorname{Pal}(v)^{-1}\operatorname{Pal}(w),$$

where the product is evaluated in the free group generated by $\{x, y\}$. (*Hint*: Using Exercise 4.8, establish that the longest palindromic suffix of $\operatorname{Pal}(w)z$ is $z\operatorname{Pal}(v)z$.)

Exercise 4.11 (Lemma 4.7). Let $\alpha_x = \mathbf{G} = (x, xy)$ and $\alpha_y = \mathbf{D} = (yx, y)$. Verify the formulae of Justin: show that for any word $w \in \{x, y\}^*$ and any $z \in \{x, y\}$,

$$\operatorname{Pal}(zw) = \alpha_z\big(\operatorname{Pal}(w)\big)z.$$

(*Hint*: Proceed by induction on $|w|$. Establish the case $|w| = 0$, then write $w = w'a$, where a is the last letter of w and consider the following two cases.

(1) If the letter a occurs in w', then write $w' = vau$, where u is a word that does not have any occurrences of a, and use Exercise 4.10 and the induction hypothesis.

(2) If the letter a does not occur in w', then show that

$$\alpha_z\big(\operatorname{Pal}(w)\big)z = \begin{cases} (\operatorname{Pal}(zw')a)^+, & \text{if } z = a, \\ (\operatorname{Pal}(zw')a)^+, & \text{if } z \neq a \end{cases} = \operatorname{Pal}(zw'a) = \operatorname{Pal}(zw)$$

using Exercise 4.10, the induction hypothesis and Exercise 4.7.
This completes the induction.)

Exercise 4.12 (Generalization of Lemma 4.7; see [Jus2005]). Let α denote the morphism defined for words in $\{x, y\}^*$ by $\alpha(x) = \mathbf{G} = (x, xy)$ and $\alpha(y) = \mathbf{D} = (yx, y)$. For any *words* $v, w \in \{x, y\}^*$,

$$\operatorname{Pal}(vw) = \alpha(v)\big(\operatorname{Pal}(w)\big)\operatorname{Pal}(v).$$

(*Hint*: Proceed by induction on $|v|$ and use Exercise 4.11.)

3. Palindromic characterization

Here we provide a converse to Theorem 4.6, namely xuy is a Christoffel word only if $u = \operatorname{Pal}(v)$ for some v in $\{x, y\}^*$. We also give a characterization of the image of Pal in terms of periods.

Lemma 4.13. *Fix an alphabet A and suppose $p \perp q$ with $p, q > 1$. Up to a permutation of A, there exists a unique word $u \in A^*$ satisfying: u has at least two distinct letters, $|u| = p + q - 2$ and u has periods p and q.*

PROOF (This proof is illustrated in the example below). Since $|u| = p+q-2$, write $u = u_1 u_2 \cdots u_{p+q-2}$, where $u_j \in A$ for $1 \leq j \leq p+q-2$. We will show that $1, 2, \ldots, p+q-2$ (the positions of the letters $u_1, u_2, \ldots, u_{p+q-2}$ in u) can be partitioned into two nonempty sets S and T such that $u_i = u_j$ if and only if i and j both belong to S or both belong to T.

Since $p \perp q$, it follows that $p \perp (p+q)$, and so p generates $\mathbb{Z}/(p+q)\mathbb{Z}$. Let G denote the Cayley graph of $\mathbb{Z}/(p+q)\mathbb{Z}$ with generator p. Consider the graph $G' = G - \{0, p+q-1\}$ obtained from G by removing the vertices 0 and $p+q-1$. Since G is connected and two vertices have been removed, G' has at most two connected components. If there is only one connected component, then 0 and $p+q-1$ are adjacent in G, and so $p+q-1$ is either $p \bmod (p+q)$ or $-p \bmod (p+q)$. The former implies that $q = 1$ and the latter implies that $p = 1$, contrary to the assumption that $p, q > 1$. Therefore, there are exactly two connected components of G'.

Suppose i and j correspond to adjacent vertices in one connected component of G'. Then $0 < i, j < p+q-1$ and either $j = i+p$ or $j = i-q$ (since $p \equiv -q \bmod (p+q)$). If $j = i+p$, then $u_j = u_{i+p} = u_i$ since u has period p; and if $j = i-q$, then $u_j = u_{i-q} = u_i$ since u has period q. It follows that $u_i = u_j$ if i and j are in the vertex set of one connected component of G'. Therefore, $u_i = a$ for all i in the vertex set of one connected component of G' and $u_j = b$ for all j in the vertex set of the other connected component of G'. Since u has at least two distinct letters, we have $a \neq b$. Thus, up to a permutation of the alphabet A, the word u is uniquely defined. □

Example. Suppose $p = 4$ and $q = 7$. The Cayley graph of $\mathbb{Z}/(4+7)\mathbb{Z}$ with generator 4 and the graph obtained by removing the vertices 0 and $4+7-1 = 10$ are shown in Figure 2. The vertex sets of the two connected components are $\{1,2,4,5,6,8,9\}$ and $\{3,7\}$. Therefore, the words $xxyxxxyxx$ and $yyxyyyxyy$ are the only words u of length $4+7-2 = 9$ with periods 4 and 7 that have at least two distinct letters.

Remark. From the proof of the lemma, one learns that the word u in question has *exactly* two distinct letters. Of course, this is already the case in the setting of interest to us, namely $A = \{x, y\}$.

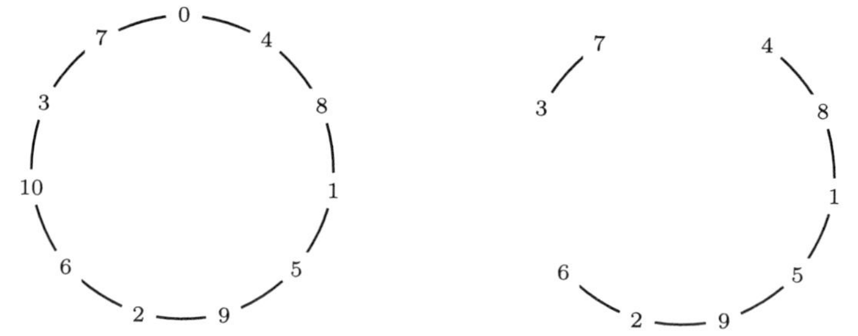

FIGURE 2. The Cayley graph of $\mathbb{Z}/(4+7)\mathbb{Z}$ with generator 4 and the connected components obtained by removing 0 and $10 = 4+7-1$.

Proposition 4.14 (de Luca, Mignosi [dLM1994]**).**

(1) $u = \mathrm{Pal}(v)$ *for some* $v \in \{x,y\}^*$ *if and only if* xuy *is a Christoffel word.*
(2) $u = \mathrm{Pal}(v)$ *for some* $v \in \{x,y\}^*$ *if and only if* u *has relatively prime periods* p *and* q *and* $|u| = p+q-2$.

PROOF OF 1. By Theorem 4.6, if $u = \mathrm{Pal}(v)$ then xuy is a Christoffel word. Conversely, let $w = xuy$ be a Christoffel word and let (w_1, w_2) be its standard factorization. Then by Corollary 3.4,

$$\begin{pmatrix} |w_1|_x & |w_2|_x \\ |w_1|_y & |w_2|_y \end{pmatrix} \in \mathrm{SL}_2(\mathbb{Z}) \cap \mathbb{N}^{2\times 2}$$

(writing $\mathbb{N}^{2\times 2}$ for the set of 2×2 matrices with nonnegative integer entries). Let N be any matrix in $\mathrm{SL}_2(\mathbb{Z}) \cap \mathbb{N}^{2\times 2}$ such that $N_{11} + N_{12} = |w|_x$ and $N_{21} + N_{22} = |w|_y$. Since $\mathrm{SL}_2(\mathbb{Z}) \cap \mathbb{N}^{2\times 2}$ is generated by the matrices $\mu(x)$ and $\mu(y)$ (see Exercise 4.2), there exists a word $v \in \{x,y\}^*$ such that $N = \mu(v)$. By Theorem 4.6, $w' = x\,\mathrm{Pal}(v)y$ is a Christoffel word and

$$\mu(v) = \begin{pmatrix} |w_1'|_x & |w_2'|_x \\ |w_1'|_y & |w_2'|_y \end{pmatrix},$$

where (w_1', w_2') is the standard factorization of w'. Since $N = \mu(v)$, it follows that $|w|_x = |w'|_x$ and $|w|_y = |w'|_y$. Thus w and w' have the same slope, and so $w = w'$ since there is a unique Christoffel word of any given slope. Since $w = xuy$ and $w' = x\,\mathrm{Pal}(v)y$, we have $u = \mathrm{Pal}(v)$. □

PROOF OF 2. By Theorem 4.6, if $u = \mathrm{Pal}(v)$, then u has relatively prime periods $|w_1|$ and $|w_2|$ and $|u| = |w_1| + |w_2| - 2$, where (w_1, w_2) is the standard factorization of the Christoffel word xuy. Conversely, suppose $u \in \{x,y\}^*$ has length $|u| = p+q-2$ and relatively prime periods $p, q > 0$. If $p = 1$ or $q = 1$, then $u = x^{p+q-2}$ or $u = y^{p+q-2}$, and in both cases $\mathrm{Pal}(u) = u$. So suppose $p, q > 1$.

Since p and q are relatively prime, there exist integers $0 < p', q' < p+q$ such that $pp' \equiv 1 \equiv qq' \bmod (p+q)$ (Exercise 1.6). We argue that $p' + q' = p + q$ and that $p' \perp q'$. Since $pp' \equiv 1 \equiv qq' \bmod (p+q)$ and $p \equiv -q \bmod (p+q)$, we have

$$p(p'+q') \equiv p'p + q'p \equiv 1 + q'p \equiv 1 - qq' \equiv 0 \bmod (p+q).$$

Therefore, $p+q$ divides $p(p'+q')$. Since p and $p+q$ are relatively prime, if follows that $p+q$ divides $p'+q'$. Since $0 < p'+q' < 2(p+q)$, we conclude that $p+q = p'+q'$. Finally, since $pp' \equiv 1 \bmod (p'+q')$, it follows that $p' \perp (p'+q')$, or $p' \perp q'$.

Let w be the Christoffel word of slope p'/q' and write $w = xu'y$ for some $u' \in \{x,y\}^*$. Then $|u'| = p' + q' - 2 = p + q - 2 = |u|$. By Part (1) of this theorem, since w is a Christoffel word, there exists a word $v \in \{x,y\}^*$ such that $u' = \mathrm{Pal}(v)$. Let (w_1, w_2) be the standard factorization of w. By Theorem 4.6, u' has periods $|w_1|$ and $|w_2|$. By Lemma 3.5, $|w_1|p' \equiv 1 \equiv |w_2|q' \bmod (p'+q')$. Since $p'+q' = p+q$, and $0 < |w_1|, |w_2| < p+q$, it follows that $|w_1| = p$ and $|w_2| = q$ because $pp' \equiv 1 \equiv qq' \bmod (p+q)$. Therefore, u' is a word of length $p+q-2$ having relatively prime periods p and q. Since such words are unique up to a permutation of the alphabet (Lemma 4.13), it follows that either $u' = u$ or $u' = \mathbf{E}(u)$. Therefore, $u = \mathrm{Pal}(v)$ or $u = \mathbf{E}(\mathrm{Pal}(v)) = \mathrm{Pal}(\mathbf{E}(v))$, where the last equality is Exercise 4.4. □

Exercise 4.13. Prove the following statements.

(a) If $p \perp q$ and w is a word of length $p + q - 1$ having periods p and q, then w is a power of a letter. (*Hint*: Use the ideas in the proof of Lemma 4.13.)

(b) (Fine–Wilf Theorem) If p and q are positive integers and w is a word of length $p + q - \gcd(p, q)$ having periods p and q, then w has period $\gcd(p, q)$.

CHAPTER 5

Primitive Elements in the Free Group F_2

In this chapter we prove that the positive primitive elements of the free group $F_2 = \langle x, y \rangle$ are conjugates of Christoffel words. We begin by recalling the relevant definitions.

1. Positive primitive elements of the free group

Let F_2 denote the free group generated by the letters x and y. Recall that every element g of F_2 has a unique representation as a reduced word over the alphabet $\{x, y, x^{-1}, y^{-1}\}$, where **reduced** means that there are no factors of the form xx^{-1}, $x^{-1}x$, yy^{-1} or $y^{-1}y$ in an expression $g = a_1 a_2 \cdots a_r$ (with $a_i \in \{x, y, x^{-1}, y^{-1}\}$). The **length** of an element $g \in F_2$, denoted by $\ell(g)$, is the length of the reduced expression for g as a word over the alphabet $\{x, y, x^{-1}, y^{-1}\}$.

A **basis** (u, v) of F_2 is a pair of elements $u, v \in F_2$ that generate F_2. A **primitive element of F_2** is an element $u \in F_2$ such that (u, v) is a basis of F_2 for some $v \in F_2$. By Exercise 5.1, any endomorphism f of F_2 is an automorphism if and only if $(f(x), f(y))$ is a basis of F_2.

Remark. We have previously used the adjective *primitive* to refer to words that are not nontrivial powers of shorter words (see Chapter 1). This will be the customary sense of the word after the present chapter as well. To avoid confusion here, we shall never omit the word *element* when referring to primitive elements of F_2.

Example. Since x and y generate F_2, the couples (x, y) and (y, x) are bases of F_2. So x and y are primitive elements of F_2.

Example. Let $u = xxy$ and $v = xxyxyxxy$. Then F_2 is generated by u and v because $u(u^{-1}vu^{-1})^{-1} = xxy(xy)^{-1} = x$ and $x^{-1}x^{-1}u = y$. Therefore, $(xxy, xxyxyxxy)$ is a basis of F_2 and xxy and $xxyxyxxy$ are primitive elements of F_2. Note that xxy is a Christoffel word and $xxyxyxxy$ is a conjugate of the Christoffel word $xxyxxyxy$.

An element $u \in F_2$ is **positive** if it is an element of the monoid $\{x, y\}^* \subseteq F_2$. Recall that $u, v \in F_2$ are called **conjugate** if there exists $g \in F_2$ such that $u = gvg^{-1}$. The following result establishes the relationship between the notions of (monoid-theoretic) conjugacy in $\{x, y\}^*$ and (group-theoretic) conjugacy in F_2.

Lemma 5.1. *Suppose u and v are positive elements of F_2. Then u and v are conjugate in F_2 if and only if u and v are conjugate in $\{x, y\}^*$.*

PROOF (Solution to Exercise 2.2). Let u and v be positive elements of F_2. That is, $u, v \in \{x, y\}^*$. Suppose u and v are conjugate in $\{x, y\}^*$. Then there exist words $w, m \in \{x,y\}^*$ such that $u = wm$ and $v = mw$. Thus $mum^{-1} = m(wm)m^{-1} = mw = v$, so u and v are conjugate in F_2.

For the converse we prove the following statement: If $h \in F_2$, then any positive elements u and v are conjugate in $\{x,y\}^*$ whenever $v = huh^{-1}$. We proceed by induction on the length of h. If the length of h is 0, then h is the identity element of F_2 and $v = u$.

Now suppose that the statement holds for all $h \in F_2$ of length less than $r > 0$. Suppose $g \in F_2$ is of length r and suppose that u and v are positive elements with $v = gug^{-1}$. Let $g = a_1 \cdots a_r$ be a reduced expression for g, where $a_i \in \{x, y, x^{-1}, y^{-1}\}$ for $1 \le i \le r$. We consider three cases.

(i) If $\ell(gu) < \ell(g) + \ell(u)$, then the first letter z of u must be $a_r^{-1} \in \{x,y\}$. Write $u = zu_1$ and $g = hz^{-1}$ for some $u_1 \in \{x,y\}^*$ and some $h \in F_2$. Then

$$v = gug^{-1} = (hz^{-1})(zu_1)(hz^{-1})^{-1} = h(u_1z)h^{-1}.$$

Since $\ell(h) < r$ and $z \in \{x,y\}$, it follows from the induction hypothesis that u_1z and v are conjugate in $\{x,y\}^*$. Since u_1z and $u = zu_1$ are conjugate in $\{x,y\}^*$, it follows that u and v are conjugate in $\{x,y\}^*$.

(ii) If $\ell(ug^{-1}) < \ell(u)+\ell(g^{-1})$, then an argument similar to that of the previous case shows that u and v are conjugate in $\{x,y\}^*$.

(iii) Finally, suppose that $\ell(gu) = \ell(g) + \ell(u)$ and $\ell(ug^{-1}) = \ell(u) + \ell(g^{-1})$. Then a reduced expression for gug^{-1} is obtained by concatenating the reduced expressions for g, u and g^{-1}. Since $u, v \in \{x,y\}^*$ and $v = gug^{-1}$, it follows that g and g^{-1} are words in $\{x,y\}^*$. Therefore, $g = 1$ and $u = v$.
This completes the induction. □

Exercise 5.1. Suppose $f\colon F_2 \to F_2$. Then $f \in \mathrm{Aut}(F_2)$ if and only if $(f(x), f(y))$ is a basis of F_2.

Exercise 5.2. Suppose r and s are nonnegative integers such that $r + s > 0$. Verify that $w = x^r y x^s$ is a primitive element of the free group $\langle x, y\rangle$ and that w is a conjugate of the Christoffel word $x^{r+s}y$.

2. Positive primitive characterization

In this section we prove a characterization of the Christoffel words based on the work of Richard P. Osborne and Heiner Zieschang [OZ1981] and Christian Kassel and Christophe Reutenauer [KR2007].

Theorem 5.2. *The words in $\{x,y\}^*$ that are conjugates of Christoffel words are exactly the positive primitive elements of the free group $F_2 = \langle x, y\rangle$.*

We begin by recalling some results about the free group F_2. There is a natural homomorphism from F_2 onto the free Abelian group $\mathbb{Z}^2$ defined by mapping the generators x and y of F_2 onto the generators $(1,0)$ and $(0,1)$ of $\mathbb{Z}^2$, respectively. This induces a map from the group $\mathrm{Aut}(F_2)$ of automorphisms of F_2 onto the group $\mathrm{Aut}(\mathbb{Z}^2) \cong \mathrm{GL}_2(\mathbb{Z})$ of automorphisms of $\mathbb{Z}^2$. The map $\mathrm{Aut}(F_2) \to \mathrm{GL}_2(\mathbb{Z})$ is given by composing an automorphism $\varphi \in \mathrm{Aut}(F_2)$ with the morphism $F_2 \to \mathbb{Z}^2$ described above. The following result of Jakob Nielsen from 1917 describes the kernel of this homomorphism. (Recall that an automorphism $\varphi\colon G \to G$ of a group G is an **inner automorphism** if there exists an $h \in G$ such that $\varphi(g) = hgh^{-1}$ for all $g \in G$.)

Theorem 5.3. *The kernel of the natural group homomorphism* $\mathrm{Aut}(\mathrm{F}_2) \to \mathrm{GL}_2(\mathbb{Z})$ *is the subgroup of inner automorphisms.*

PROOF. A proof of this classical result from combinatorial group theory can be found in either [LS2001, Chapter I, Proposition 4.5] or [MKS2004]. □

The following result characterizes pairs of generators of the free Abelian group $\mathbb{Z}^2$.

Lemma 5.4. (a, b) *and* (c, d) *generate* $\mathbb{Z}^2$ *if and only if* $|ad - bc| = 1$.

PROOF. Suppose $|ad - bc| = 1$. Then $(1, 0)$ and $(0, 1)$ are in the span of (a, b) and (c, d). Indeed, $d(a, b) - b(c, d) = (ad - bc, 0) = \pm(1, 0)$ and $a(c, d) - c(a, b) = \pm(0, 1)$. Thus (a, b) and (c, d) generate $\mathbb{Z}^2$.

Conversely, suppose (a, b) and (c, d) generate $\mathbb{Z}^2$. Then the matrix equation

$$\begin{pmatrix} a & c \\ b & d \end{pmatrix} \vec{x} = \vec{b}$$

has a unique solution $\vec{x} \in \mathbb{Z}^2$ for all vectors $\vec{b} \in \mathbb{Z}^2$. For $\vec{b} = (0, 1)^T$, we have

$$\vec{x} = \begin{pmatrix} a & c \\ b & d \end{pmatrix}^{-1} \begin{pmatrix} 0 \\ 1 \end{pmatrix} = \frac{1}{ad - bc} \begin{pmatrix} d & -c \\ -b & a \end{pmatrix} \begin{pmatrix} 0 \\ 1 \end{pmatrix} = \frac{1}{ad - bc} \begin{pmatrix} -c \\ a \end{pmatrix} \in \mathbb{Z}^2.$$

It follows that $|ad - bc|$ divides $|a|$ and $|c|$. Since (a, b) and (c, d) generate $\mathbb{Z}^2$, there exists $i, j \in \mathbb{Z}$ such that $i(a, b) + j(c, d) = (1, 0)$. In particular, $ia + jc = 1$, from which it follows that $|a| \perp |c|$ (see *Bézout's Lemma* in Exercise 3.3). Since $|ad - bc|$ divides $|a|$ and $|c|$, and since $|a| \perp |c|$, we have $|ad - bc| = 1$. □

PROOF OF THEOREM 5.2. Suppose w is a Christoffel word and let (u, v) denote its standard factorization. By Theorem 3.6, the couple (u, v) is in the Christoffel tree. Hence, $(w, v) = (uv, v)$ is in the Christoffel tree as well. Since the root (x, y) of the Christoffel tree is a basis of F_2 and since (u, uv) and (uv, v) are bases of F_2 whenever (u, v) is a basis of F_2, it follows that each couple (u, v) in the Christoffel tree is a basis of F_2. In particular, (w, v) is a basis of F_2, thus w is a positive primitive element of F_2.

Now suppose $w' \in \{x, y\}^*$ is a conjugate of a Christoffel word w. By Lemma 5.1, there exists $u \in \{x, y\}^*$ such $w' = u^{-1}wu$. Since w is a primitive element of F_2, there exists $v \in \mathrm{F}_2$ such that (w, v) is a basis of F_2. Thus $(w', u^{-1}vu) = (u^{-1}wu, u^{-1}vu)$ is a basis of F_2 and w' is a positive primitive element of F_2 as well.

Conversely, suppose w is a positive primitive element of F_2. Then there exists $w' \in \mathrm{F}_2$ such that (w, w') is a basis of F_2. Therefore, $g = (w, w')$ is an automorphism of F_2 (by Exercise 5.1). Our proof will construct an automorphism $f = (u, v)$ of F_2 with u and v Christoffel words and show that $g = (w, w')$ is conjugate (in the group-theoretic sense) to $f = (u, v)$. This will imply that w is a conjugate of u (and w' is a conjugate of v) in $\{x, y\}^*$. We begin by analyzing the image of g in $\mathbb{Z}^2$.

Define a group homomorphism $\mathrm{F}_2 \to \mathbb{Z}^2$ by $x \mapsto (1, 0)$ and $y \mapsto (0, 1)$ and let (a, b) and (c, d) denote the images of w and w', respectively, under this homomorphism. Note that (a, b) and (c, d) generate $\mathbb{Z}^2$ because (w, w') is a basis of the free group F_2. Therefore, by the previous lemma, $ad - bc = \pm 1$.

Note that a and b are nonnegative since w is a positive element of F_2. Let us analyze the possibilities for small values of a and b. Both cannot be zero since $ad - bc = \pm 1$. If $a = 0$, then $\pm 1 = ad - bc = -bc$, which implies that $b = 1$.

Thus $w = y$, which is a Christoffel word. If $b = 0$, then $ad - bc = \pm 1$ implies that $a = 1$. So $w = x$, which is a Christoffel word. Hence, we suppose a and b are positive integers. Fix $n \geq 1$. A direct computation (see Exercise 5.2) reveals that all $w \in F_2$ with $(a, b) = (n, 1)$, respectively $(a, b) = (1, n)$, are positive primitive elements of F_2 and are conjugate to the Christoffel word $x^n y$, respectively $y^n x$. Hence, we further suppose $a, b \geq 2$.

If $c < 0$ and $d \geq 0$ or if $c \geq 0$ and $d < 0$, then $|ad - bc| \geq 2$, contradicting the fact that $|ad - bc| = 1$. Therefore, c and d are both nonnegative or both nonpositive. If c and d are both nonpositive, then replace w' with $(w')^{-1}$; thus we can assume that c and d are nonnegative. Finally, we assume $ad - bc = 1$ (otherwise, swap w and w' in what follows). In summary, we have a basis (and automorphism) $g = (w, w')$ of F_2 such that: the images (a, b) and (c, d) of w and w' generate $\mathbb{Z}^2$; the points (a, b) and (c, d) lie in the first quadrant; and $ad - bc = 1$. Hence,

$$\begin{pmatrix} a & c \\ b & d \end{pmatrix} \in \mathrm{SL}_2(\mathbb{Z}) \cap \mathbb{N}^{2\times 2}.$$

Define a semigroup morphism $M \colon \{\mathbf{G}, \widetilde{\mathbf{D}}\}^* \to \mathrm{SL}_2(\mathbb{Z}) \cap \mathbb{N}^{2\times 2}$ by

$$M(\mathbf{G}) = \begin{pmatrix} 1 & 1 \\ 0 & 1 \end{pmatrix} \quad \text{and} \quad M(\widetilde{\mathbf{D}}) = \begin{pmatrix} 1 & 0 \\ 1 & 1 \end{pmatrix}.$$

Since the monoid $\mathrm{SL}_2(\mathbb{Z}) \cap \mathbb{N}^{2\times 2}$ is generated by the 2×2 matrices $M(\mathbf{G})$ and $M(\widetilde{\mathbf{D}})$ (see Exercise 4.2), there exists an endomorphism f of $\{x, y\}^*$ such that f is a composition of the morphisms $\mathbf{G}$ and $\widetilde{\mathbf{D}}$, and

$$M(f) = \begin{pmatrix} a & c \\ b & d \end{pmatrix}.$$

Since f is a composition of the morphisms $\mathbf{G}$ and $\widetilde{\mathbf{D}}$, the couple $(f(x), f(y))$ is an element of the Christoffel tree. Therefore, $(f(x), f(y))$ is a basis of F_2 by the first paragraph of this proof. In particular, f is an automorphism of F_2 and the elements $u = f(x)$ and $v = f(y)$ are Christoffel words.

We now compute the composition of f with the natural morphism $F_2 \to \mathbb{Z}^2$ defined above. Note that we need only compute the images of $f(x)$ and $f(y)$ since they generate $\mathbb{Z}^2$. Note that the image of $f(x)$ is $(|f(x)|_x, |f(x)|_y)$ and the image of $f(y)$ is $(|f(y)|_x, |f(y)|_y)$. Since $\mathbf{G}$ replaces each letter of a word m with a word having exactly one occurrence of x, we have $|\mathbf{G}(m)|_x = |m|$. Since $\mathbf{G}$ replaces y with xy, we have $|\mathbf{G}(m)|_y = |m|_y$. Equivalently, for any word $m \in \{x, y\}^*$,

$$\begin{pmatrix} |\mathbf{G}(m)|_x \\ |\mathbf{G}(m)|_y \end{pmatrix} = \begin{pmatrix} |m| \\ |m|_y \end{pmatrix} = \begin{pmatrix} 1 & 1 \\ 0 & 1 \end{pmatrix} \begin{pmatrix} |m|_x \\ |m|_y \end{pmatrix} = M(\mathbf{G}) \begin{pmatrix} |m|_x \\ |m|_y \end{pmatrix}.$$

Similarly,

$$\begin{pmatrix} |\widetilde{\mathbf{D}}(m)|_x \\ |\widetilde{\mathbf{D}}(m)|_y \end{pmatrix} = \begin{pmatrix} |m|_x \\ |m| \end{pmatrix} = \begin{pmatrix} 1 & 0 \\ 1 & 1 \end{pmatrix} \begin{pmatrix} |m|_x \\ |m|_y \end{pmatrix} = M(\widetilde{\mathbf{D}}) \begin{pmatrix} |m|_x \\ |m|_y \end{pmatrix}.$$

Since f is a composition of the morphisms $\mathbf{G}$ and $\widetilde{\mathbf{D}}$, these two identities imply that the number of occurrences of the letters x and y in the word $f(m)$ is given by

$$\begin{pmatrix} |f(m)|_x \\ |f(m)|_y \end{pmatrix} = M(f) \begin{pmatrix} |m|_x \\ |m|_y \end{pmatrix} = \begin{pmatrix} a & c \\ b & d \end{pmatrix} \begin{pmatrix} |m|_x \\ |m|_y \end{pmatrix}.$$

In particular, $(|f(x)|_x, |f(x)|_y) = (a, b)$ and $(|f(y)|_x, |f(y)|_y) = (c, d)$. Therefore, the image of u in $\mathbb{Z}^2$ is (a, b) and the image of v in $\mathbb{Z}^2$ is (c, d).

In summary, $g = (w, w')$ and $f = (u, v)$ are two automorphisms of F_2 that give the same morphism of $\mathbb{Z}^2$ after composing with the map $\mathrm{F}_2 \to \mathbb{Z}^2$. By Theorem 5.3, the morphism $f^{-1}g$ is an inner automorphism of F_2. We conclude that there exists $z \in \mathrm{F}_2$ such that $f^{-1}g(m) = zmz^{-1}$ for all $m \in \mathrm{F}_2$. Applying f to each side of this equality, we get $g(m) = f(z)f(m)f(z)^{-1}$. In particular, $w = g(x)$ is conjugate in F_2 to the Christoffel word $u = f(x)$ and $w' = g(y)$ is conjugate in F_2 to the Christoffel word $v = f(y)$. Then Lemma 5.1 yields that w and w' are conjugates of Christoffel words in the monoidal sense. □

By refining the method of the above proof one can obtain the following result relating the Christoffel morphisms with the automorphisms of the free group F_2. Recall that an element $w \in \mathrm{F}_2$ is positive if $w \in \{x, y\}^* \subseteq \mathrm{F}_2$. An endomorphism f of F_2 is a **positive morphism** if both $f(x)$ and $f(y)$ are positive.

Theorem 5.5 (Wen, Wen [WW1994]). *The Christoffel morphisms of* $\{x, y\}^*$ *are exactly the positive morphisms of the free group* $\langle x, y\rangle$.

CHAPTER 6

Characterizations

By now we have presented several characterizations of Christoffel words—discretization of line segments, Cayley graphs of cyclic groups, palindromization and the positive primitive elements of F_2. In this chapter we present a few more, beginning with one that we have already met in passing.

If w is a (lower) Christoffel word, then by definition it looks like xuy for some $u \in \{x, y\}^*$. After Lemma 2.7 and Proposition 4.2, we moreover know that w is a conjugate of the word yux. The converse also holds. See Figure 1.

Theorem 6.1 (Pirillo [Pir1999]**).** *Given $u \in \{x, y\}^*$, xuy is a Christoffel word if and only if xuy and yux are conjugate.*

1. The Burrows–Wheeler transform

Suppose w is a word over an ordered alphabet[1] A. For example, take w to be the Christoffel word $xxyxy$ or the word $abraca$. Write w and all of its distinct conjugates as rows in a matrix, listed in lexicographic order.[2] The **Burrows–Wheeler transform** $\mathrm{BWT}(w)$ of w is the last column of this matrix. See Figure 2.

This is not quite the map introduced by Michael Burrows and David J. Wheeler [BW1994]. Their map, call it BWT^+, maps a word w to the pair $(\mathrm{BWT}(w), k)$, whenever w appears as the kth row of the Burrows–Wheeler matrix. For example, $\mathrm{BWT}^+(bracaa) = (caraab, 4)$ (see Figure 2). This augmented map turns out to be injective on A^* (Exercise 6.2). It was introduced as a scheme for lossless data compression, and as Giovanni Manzini shows, BWT^+ is very effective at its intended purpose [Man2001].

Alternative to introducing BWT^+, we could restrict our attention to those words that are lexicographically least among their conjugates. These are the so-called **Lyndon words**. Evidently, BWT becomes injective under this restriction. More interestingly, it becomes a special case of a mapping introduced by Ira M. Gessel and Christophe Reutenauer in the study of descent sets of permutations [GR1993]. See also [CDP2005].

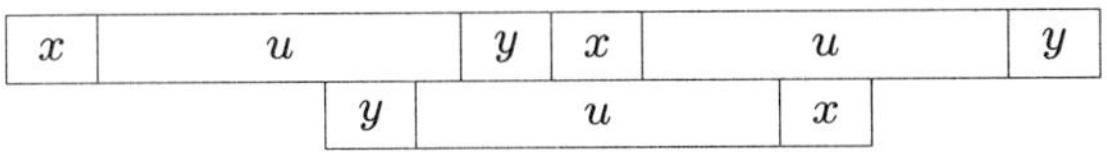

FIGURE 1. The word yux is a conjugate of xuy if and only if xuy is a Christoffel word.

---

[1] We use the inherited ordering for all subsets A of $\{a < b < c < \cdots < z\}$. In particular, x precedes y in our favourite alphabet $\{x, y\}$.

[2] The usual dictionary ordering, where *aardvark* comes before *ant* comes before *anthill*.

$$
\begin{array}{c}
\underline{x\,x\,y\,x\,y} \\
\downarrow \\
x\;x\;y\;x\;\mathbf{y} \\
x\;y\;x\;x\;\mathbf{y} \\
x\;y\;x\;y\;\mathbf{x} \\
y\;x\;x\;y\;\mathbf{x} \\
y\;x\;y\;x\;\mathbf{x}
\end{array}
\qquad
\begin{array}{c}
\underline{a\,b\,r\,a\,c\,a} \\
\downarrow \\
a\;a\;b\;r\;a\;\mathbf{c} \\
a\;b\;r\;a\;c\;\mathbf{a} \\
a\;c\;a\;a\;b\;\mathbf{r} \\
b\;r\;a\;c\;a\;\mathbf{a} \\
c\;a\;a\;b\;r\;\mathbf{a} \\
r\;a\;c\;a\;a\;\mathbf{b}
\end{array}
$$

FIGURE 2. Instances of the Burrows–Wheeler transform. The transforms $xxyxy \mapsto yyxxx$ and $abraca \mapsto caraab$ are obtained by reading the last columns of the matrices above.

When $w = xuy$ is a (lower) Christoffel word, it happens that w appears as the first row of its Burrows–Wheeler matrix (i.e., Christoffel words are Lyndon words). This and other interesting properties are illustrated in Figure 3. The first two are explained in Exercise 6.3. The third is explained by our next characterization of Christoffel words.

Theorem 6.2 (Mantaci, Restivo, Sciortino [MRS2003]**).** *A word* $w \in \{x, y\}^*$ *is a conjugate of a Christoffel word if and only if* $\mathrm{BWT}(w)$ *takes the form* $y^q x^p$, *where* $p \perp q$.

Exercise 6.1. Prove Theorem 6.1. (*Hint*: This is a consequence of the Fine–Wilf Theorem; see Exercise 4.13.)

Exercise 6.2. The Burrows–Wheeler transform BWT is injective on Lyndon words (cf. [CDP2005] for more details):

Given a Lyndon word $w = a_1 a_2 \cdots a_n$, let $b_1 b_2 \cdots b_n$ and $c_1 c_2 \cdots c_n$ denote the first and last columns, respectively, of the Burrows–Wheeler matrix.

(a) Define a permutation $\sigma \in \mathfrak{S}_n$ by putting $\sigma(i) = j$ if the jth row of the Burrows–Wheeler matrix is $a_i a_{i+1} \cdots a_{i-1}$. Verify that $a_i = b_{\sigma(i)}$.

(b) Define a permutation $\pi \in \mathfrak{S}_n$ by the n-cycle $\pi = \big(\sigma(1)\sigma(2)\ldots\sigma(n)\big)$. Verify that $b_i = c_{\pi(i)}$.

(c) Let w' be a Lyndon word with $\mathrm{BWT}(w') = \mathrm{BWT}(w) = c_1 c_2 \cdots c_n$. Deduce that $w = w'$.

$$
\begin{array}{c|cc|cc}
\cline{2-3}
x & x & y & x & y \\
x & y & x & x & y \\
\cline{2-3}
x & y & x & y & x \\
\cline{3-4}
y & x & \multicolumn{1}{|c}{x} & \multicolumn{1}{c|}{y} & x \\
y & x & \multicolumn{1}{|c}{y} & \multicolumn{1}{c|}{x} & x \\
\cline{3-4}
\end{array}
$$

FIGURE 3. The Burrows–Wheeler matrix for the Christoffel word of slope q/p possesses three interesting properties: (i) the lower- and upper-Christoffel words comprise the first and last rows, respectively; (ii) any two consecutive rows differ in two consecutive positions; (iii) the last column takes the form $y^q x^p$.

Exercise 6.3 (Property (ii) of Figure 3). Given the Christoffel word w of slope q/p, let w_t denote the conjugate of w obtained by reading $|w|$ letters along the Christoffel path of ww, starting at the lattice point C labelled t/p. For example, $w_0 = w$ and w_{p+q-1} is the upper Christoffel word of slope q/p (see the proof of Lemma 2.7). Let $n_t(k)$ denote the numerator of the label k steps after C along the Christoffel path for w_t. That is, $n_t(k) = t + kq \bmod (p+q)$.

(a) If two successive conjugates w_{t-1} and w_t have longest common prefix u, then $w_{t-1} = uxyv$ and $w_t = uyxv'$. In particular, $n_t(0) - n_{t-1}(0) = 1$ and $n_t(|u| + 2) - n_{t-1}(|u|+2) = 1$.

(b) In general, one has $n_{t-1}(k) \bmod (p+q) < n_t(k) \bmod (p+q)$. There is precisely one instance of $0 \le k < p+q$ with $n_t(k) \bmod (p+q) < n_{t-1}(k) \bmod (p+q)$.

(c) In the factorizations $uxyv$ and $uyxv'$ of part (a), one has $v = v'$.

(d) The $(t+1)$st row of the Burrows–Wheeler matrix for w is w_t.

2. Balanced$_1$ Lyndon words

We next introduce an important class of words first defined by Marston Morse and Gustav A. Hedlund in 1940. A word $w \in \{x, y\}^*$ is **balanced$_1$** if for each pair u, v of factors of w of equal length, one has

$$\big|\,|u|_x - |v|_x\big| \le 1, \quad \text{or equivalently} \quad \big|\,|u|_y - |v|_y\big| \le 1.$$

Serge Dulucq and Dominique Gouyou-Beauchamps [DG1990] have shown that the set of balanced$_1$ words is exactly the set of factors of Christoffel words, or equivalently of Sturmian words (cf. [Lot2002, Chapter 2]). The characterization we seek is the following.

Theorem 6.3 (de Luca, Mignosi [dLM1994]). *A word xuy is a Christoffel word if and only if xux, xuy, yux, and yuy are balanced$_1$.*

Alternatively, one may replace the extra "balanced$_1$" checks with a single "Lyndon" check.

Theorem 6.4 (Berstel, de Luca [BdL1997]). *A word w is a Christoffel word if and only if it is a balanced$_1$ Lyndon word.*

3. Balanced$_2$ Lyndon words

There is another notion of "balanced" that may be given to Lyndon words. Before defining it, we need to introduce a fundamental result about Lyndon words. Recall that

> *a word w is a **Lyndon word** if and only if for all nontrivial factorizations $w = (u, v)$, $w < vu$ in the lexicographic order.*

Note that we did not allow $w \le vu$. Otherwise stated, we demand that Lyndon words are primitive. We thus have an equivalent definition: w is a **Lyndon word** if $w < v$ for all proper suffixes v of w. If we choose v to be the lexicographically least suffix of w, a surprising thing happens (cf. [Lot1997, Chapter 5]).

Proposition 6.5 (Chen, Fox, Lyndon [CFL1958], Duval [Duv1983]). *If $w = uv$ is a Lyndon word with v its lexicographically least proper suffix, then u and v are also Lyndon words and $u < v$.*

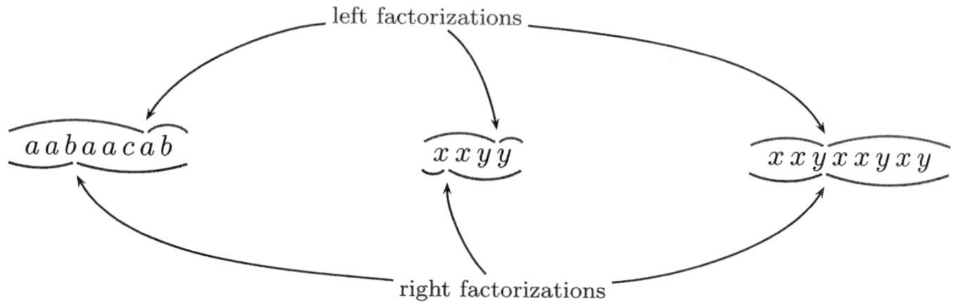

FIGURE 4. The left and right factorizations of three Lyndon words. Only $xxyxxyxy$, the Christoffel word of slope $3/5$, is seen to be a balanced$_2$ Lyndon word.

This is the standard factorization of a Lyndon word, which we call the **right factorization** in what follows. It happens that v is simultaneously the longest proper suffix that is Lyndon, which brings us to an alternative **left factorization** due to Anatolii I. Shirshov [Šir1962] and Xavier G. Viennot [Vie1978].

Proposition 6.6. *If $w = uv$ is a Lyndon word with u a proper Lyndon prefix of maximal length, then v is also a Lyndon word and $u < v$.*

The left factorization and right factorization of a Lyndon word sometimes coincide. This led Guy Melançon [Mel1999] to introduce a second, recursive, notion of *balanced*: call a Lyndon word w **balanced$_2$** if w is a letter or there is a factorization $w = (u, v)$ that is both a left and right factorization with the further property that u and v are balanced$_2$.

Example. The reader may check that $aabaacab$, $xxyy$, and $xxyxxyxy$ are Lyndon words. Among these, only $xxyxxyxy$ is balanced$_2$. See Figure 4.

Theorem 6.7 (Melançon [Mel1999]**).** *A word $w \in \{x, y\}^*$ is a Christoffel word if and only if it is a balanced$_2$ Lyndon word.*

4. Circular words

Many of the results in this and the preceding chapters deal with conjugates of Christoffel words, but do not distinguish one conjugate from another. Such results are perhaps better described in terms of *circular words:* the conjugacy class of a word w is called a **circular word** and is denoted by (w). Our next characterization makes explicit mention of these words.

Theorem 6.8 (Borel, Reutenauer [BR2006]**).** *The following are equivalent for a word $w \in \{x, y\}^*$ of length $n \geq 2$:*

(i) *w is a conjugate of a Christoffel word;*
(ii) *the circular word (w) has $k + 1$ factors of length k for $k = 0, 1, \ldots, n - 1$;*
(iii) *w is a primitive word and (w) has $n - 1$ factors of length $n - 2$.*

Example. Take $w = yxxxyxx$, which is a conjugate of the Christoffel word of slope $2/5$. We list the distinct factors of each length in Figure 5.

A brief proof of Theorem 6.8 affords us the opportunity to introduce several additional results from the theory of Sturmian words.

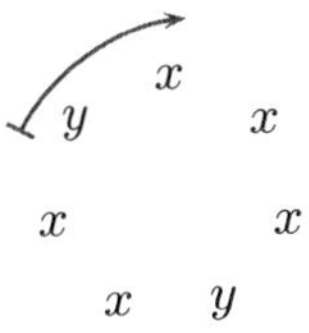

ℓ	distinct factors of length ℓ
1	y x
2	yx xx xy
3	yxx xxx xxy xyx
4	$yxxx$ $xxxy$ $xxyx$ $xyxx$ $yxxy$
5	$yxxxy$ $xxxyx$ $xxyxx$ $xyxxy$ $yxxxy$ $xyxxx$
6	$yxxxyx$ $xxxyxx$ $xxyxxy$ $xyxxyx$ $yxxyxx$ $xxyxxx$ $xyxxxy$

FIGURE 5. For Christoffel words w, there are $\ell+1$ distinct factors of length $\ell = 1, 2, \ldots$ in the circular words (w).

Theorem 6.9 (folklore). *A word $w \in \{x,y\}^*$ is a conjugate of a Christoffel word if and only if w and all of its conjugates are primitive and balanced$_1$.*

Remarks.

(1) The requirement that all conjugates be balanced$_1$ is essential here. For example, $xyyx$ is balanced$_1$ but is certainly not conjugate to a Christoffel word.

(2) Antonio Restivo attributes this theorem to Oliver Jenkinson and Luca Q. Zamboni [JZ2004] when he speaks about the Burrows – Wheeler transform. He calls a word *strongly balanced* if it satisfies the conditions of the theorem. We indicate below a proof suggested by Geneviève Paquin.

PROOF. The forward direction is essentially [Lot2002, Proposition 2.1.10]. First, conjugates of primitive words are primitive, so we need only analyze the balanced$_1$ implication. This follows from a geometric argument. Suppose w is a conjugate of the Christoffel word of slope b/a. If some conjugate of w is not balanced$_1$, then there are two factors u and v of ww of the same length with $|u|_y$ and $|v|_y$ differing by at least 2. On the other hand, u and v follow the line of slope b/a, so the quantities $|u|_y$ and $|v|_y$ can differ by at most 1.

The reverse direction follows from Theorem 6.4. Indeed, if w and all of its conjugates are primitive and balanced$_1$, then the lexicographically least conjugate w' is a Lyndon word (since it is primitive). Hence w' is a Christoffel word and w is a conjugate of a Christoffel word. □

A sequence (infinite word) $w = a_0a_1a_2\cdots$ over an alphabet A is called **ultimately periodic** if w can be factored as $w = uv^\infty$ for some finite words $u, v \in A^*$. We do not assume $u \neq \epsilon$ so periodic words are ultimately periodic. If w is not ultimately periodic, we say that w is **aperiodic**.

Theorem 6.10 (Morse, Hedlund [MH1940]). *If a sequence in $\{x,y\}^{\mathbb{N}}$ is aperiodic, then it is balanced$_1$ if and only if it has exactly $k+1$ factors of length k for all $k \geq 0$.*

Remark. The factor complexity described above is often taken as the definition of **Sturmian word**. In the proof sketch below, we will only need the trivial part of the forward direction, cf. [Lot2002, Proposition 2.1.2]: *a balanced$_1$ sequence in $\{x,y\}^{\mathbb{N}}$ has at most $k+1$ factors of length k for all $k \geq 0$.*

Theorem 6.11 (Coven, Hedlund [CH1973]). *A sequence in $\{x,y\}^{\mathbb{N}}$ is aperiodic if and only if it has at least $k+1$ factors of length k for all $k \geq 0$.*

Remark. Periodic sequences with balanced$_1$ factors are sometimes excluded from the definition of Sturmian words in the literature. Ultimately periodic (but not periodic) sequences are called "skew-words" by Morse and Hedlund [MH1940]. In the sketch below, we will actually use a "circular" version of this theorem, cf. [BR2006, Lemma 4.1]: *a word* w *is primitive if and only if* (w) *has at least* $k+1$ *factors of length* k *for all* $0 \leq k < |w|$.

PROOF OF THEOREM 6.8. (i) $\Rightarrow$ (ii) $\Rightarrow$ (iii): Suppose w is a conjugate of a Christoffel word of length n. Then w and its conjugates are primitive and balanced$_1$ (Theorem 6.9). Hence, (w) has at most $k+1$ factors of length k and at least $k+1$ factors of length k for all $0 \leq k < n$ (by the two remarks above). In particular, w is primitive and (w) has exactly $n-1$ factors of length $n-2$.

(iii) $\Rightarrow$ (i): Suppose w is primitive and (w) has exactly $n-1$ factors of length $n-2$. Then (w) has at least $k+1$ factors of length k for all $0 \leq k < n$ (by the remark following Theorem 6.11). This implies the existence of a special factor u of length $n-2$ with $(w) = (xuy) = (yux)$ (Exercise 6.4). But then either xuy or yux is a Christoffel word (Theorem 6.1). That is, w is a conjugate of a Christoffel word. □

Exercise 6.4. If w is a word of length n and (w) has at least n factors of length $n-1$, then there is a factor u of (w) of length $n-2$ so that $(w) = (xuy) = (yux)$. In particular, w is a conjugate of a Christoffel word, by Theorem 6.1. (*Hint*: The u that you seek is the unique factor of length $n-2$ that may be extended to the left in two ways (and to the right in two ways) to a factor of length $n-1$.)

5. Periodic phenomena

Our final characterization of Christoffel words is another folklore result that says that the superposition of two periodic phenomena gives rise to all the Christoffel words. It is anticipated in the original papers on the subject [Chr1875, Smi1876].

Theorem 6.12 (Superposition of two periodic phenomena). *Suppose* p *and* q *are positive integers and* $p \perp q$. *Set* $P = \{ip : 0 < i < q\}$ *and* $Q = \{jq : 0 < j < p\}$. *Write* $P \cup Q$ *as* $\{a_1, a_2, \ldots, a_n\}$, *where* $a_1 < a_2 < \cdots < a_n$ *and* $n = p+q-2$. *Then the word* $xw_1w_2 \cdots w_ny$, *where* $w_i = x$ *if* $a_i \in P$ *and* $w_i = y$ *if* $a_i \in Q$, *is the Christoffel word of slope* p/q.

The proof is left as an exercise.

Examples.

(1) Let $p = 4$ and $q = 7$. Then p and q are relatively prime, $P = \{4, 8, 12, 16, 20, 24\}$ and $Q = \{7, 14, 21\}$. The superposition of P and Q is given below with the elements of Q in boldface.

	4	**7**	8	12	**14**	16	20	**21**	24	
x	x	y	x	x	y	x	x	y	x	y

Thus we obtain the Christoffel word $xxyxxyxxyxy$ of slope $4/7$.

(2) Another example (of length 99) appears implicitly in the following passage from *The Brooklyn Follies* by Paul Auster:

> This was the truth of the world, she told her father at breakfast that morning, and in order to get a grip on that truth, she had decided to spend the day sitting in the rocking chair in her room, shouting out the

word *rejoice* every forty-one seconds and the word *grieve* every fifty-eight seconds ... [Aus2006, p. 50].

(3) Our final example is arguably one of the very first examples. Caroline Series suggests [Ser1985], tongue-in-cheek, that the earliest astronomers were acquainted with Theorem 6.12. Indeed, in Babylonian calendars from 490 B.C., one finds extremely accurate approximations such as, "19 years = 235 lunar months." Surely the original calculation was written down in tally form, "month, month, ..., month, year, month, month, ..., month, year, ...," forming one of the first recorded Christoffel words.

CHAPTER 7

Continued Fractions

This chapter exhibits relationships between the Christoffel tree, continued fractions and the Stern–Brocot tree of positive rational numbers. See the articles [BL1993, GKP1994, BdL1997] for further information, and for a geometric approach to continued fractions, see [Dav1992].

1. Continued fractions

Suppose $\alpha \in \mathbb{R}$. The (simple) **continued fraction representation** of α is the sequence of integers $a_0, a_1, a_2, \ldots$ constructed recursively as follows: let

$$\beta_0 = \alpha \quad \text{and} \quad a_0 = \lfloor \beta_0 \rfloor;$$

if $i > 0$ and $a_{i-1} \neq \beta_{i-1}$, then let

$$\beta_i = \frac{1}{\beta_{i-1} - a_{i-1}} \quad \text{and} \quad a_i = \lfloor \beta_i \rfloor;$$

if $i > 0$ and $a_{i-1} = \beta_{i-1}$, then the recursion terminates. The continued fraction representation of α is commonly denoted by

$$\alpha = a_0 + \cfrac{1}{a_1 + \cfrac{1}{a_2 + \cfrac{1}{a_3 + \cfrac{1}{\ddots}}}}$$

or more compactly by $\alpha = [a_0, a_1, a_2, a_3, \ldots]$.

Example. Let $\alpha = 10/23$. Then

$$\begin{aligned}
\beta_0 &= \frac{10}{23}, & a_0 &= \lfloor \beta_0 \rfloor = 0;\\
\beta_1 &= \frac{1}{\beta_0 - a_0} = \frac{23}{10} & a_1 &= \lfloor \beta_1 \rfloor = 2;\\
\beta_2 &= \frac{1}{\beta_1 - a_1} = \frac{10}{3}, & a_2 &= \lfloor \beta_2 \rfloor = 3;\\
\beta_3 &= \frac{1}{\beta_2 - a_2} = 3, & a_3 &= \lfloor \beta_3 \rfloor = 3.
\end{aligned}$$

Since $\beta_3 = a_3$, the process ends and the continued fraction representation of 10/23 is $[\mathbf{0}, \mathbf{2}, \mathbf{3}, \mathbf{3}]$, or

$$\frac{10}{23} = \mathbf{0} + \cfrac{1}{\mathbf{2} + \cfrac{1}{\mathbf{3} + \cfrac{1}{\mathbf{3}}}}.$$

Note that, by construction, the integers $a_1, a_2, \ldots$ are positive. The continued fraction representation of $\alpha \in \mathbb{R}$ is finite (that is, the above process terminates) if and only if α is a rational number.

If $[a_0, a_1, a_2, \ldots, a_i, \ldots]$ is the continued fraction representation of α, then the ith **continuant** of α, for $i > 0$, is the rational number with continued fraction representation $[a_0, a_1, \ldots, a_{i-1}, a_i]$.

Example. From the previous example, the continued fraction representation of 10/23 is given by the sequence $[0, 2, 3, 3]$. Therefore, the continuants of 10/23 are

$$0 + \frac{1}{2} = \mathbf{\frac{1}{2}}, \quad 0 + \cfrac{1}{2 + \cfrac{1}{3}} = \mathbf{\frac{3}{7}}, \quad 0 + \cfrac{1}{2 + \cfrac{1}{3 + \cfrac{1}{3}}} = \mathbf{\frac{10}{23}}. \tag{7.1}$$

2. Continued fractions and Christoffel words

In his 1876 note [Smi1876], Henry J. Smith showed that the sequences obtained from periodic phenomena in Chapter 5 can be obtained from the continued fraction representation of the associated rational number, and vice versa. He effectively proved the following characterization of Christoffel words.

Theorem 7.2 (Smith [Smi1876]). *A word $w = xuy$ is a Christoffel word if and only if uyx or uxy is equal to s_n, where s_n is defined recursively by $s_{-1} = x$, $s_0 = y$, and $s_{n+1} = s_n^{c_n} s_{n-1}$ for $n \geq 0$, where $[c_0, c_1, \ldots]$ is the continued fraction representation of $|w|_y/|w|_x$.*

Examples.

(1) The continued fraction representation of 4/7 is $[0, 1, 1, 3]$. Thus,

$$\begin{aligned} s_1 &= s_0^0 s_{-1} = x, \\ s_2 &= s_1^1 s_0 = xy, \\ s_3 &= s_2^1 s_1 = xyx, \\ s_4 &= s_3^3 s_2 = (xyxxyxxyx)xy, \end{aligned}$$

and indeed, $x(xyxxyxxyx)y$ is the the Christoffel word of slope 4/7.

(2) The continued fraction representation of 2/5 is $[0, 2, 2]$. Hence,

$$\begin{aligned} s_1 &= s_0^0 s_{-1} = x, \\ s_2 &= s_1^2 s_0 = xxy, \\ s_3 &= s_2^2 s_1 = (xxyxx)yx, \end{aligned}$$

and $x(xxyxx)y$ is the Christoffel word of slope 2/5.

Smith's theorem gives a method to obtain the continued fraction representation $[c_0, c_1, c_2, \ldots]$ of any positive rational number α by considering the Christoffel word xuy of slope α. Let v be uxy (or uyx if uxy does not work). The integers c_i are determined inductively as follows. Let c_0 be the highest power of y that is a prefix of v. Suppose that $c_0, c_1, \ldots, c_i$ and $s_0, s_1, \ldots, s_{i+1}$ have been constructed. Then c_{i+1} is the largest integer n such that $s_{i+1}^n s_i$ is a prefix of v. We illustrate this procedure with the following example.

Example. The Christoffel word of slope 4/7 is $x(xyxxyxxyx)y$. Let v be the word $xyxxyxxyxxy$. Then $c_0 = 0$ since v does not begin with y. Since $s_1 s_0 = xy$ is a prefix of v, but $s_1^2 s_0 = x^2 y$ is not, we have $c_1 = 1$. Since $s_2 s_1 = xyx$ is a prefix of v while $s_2^2 s_1 = (xy)^2 x$ is not, we have $c_2 = 1$. Finally, $c_3 = 3$ since $v = (xyx)^3 xy = (s_3)^3 s_2$.

In [Smi1876], one also finds a geometric method of obtaining the continuants of real numbers (see Section 20, loc. cit.). We now explain this method for rational numbers $b/a > 0$. To find the continuants for an irrational number α, simply replace the line segment from $(0,0)$ to (a,b) in what follows by the ray of slope α.

Consider the subpath of the lower Christoffel path from $(0,0)$ to (a,b) beginning at $(1,0)$ and ending at (a,b). For example, if $(a,b) = (23,10)$, then this subpath is the lower path depicted in Figure 1. The convex hull of this subpath determines a sequence of integer points beginning with $(1,0)$ and ending with (a,b) by following the upper boundary from left to right. For $(a,b) = (23,10)$, we have the sequence $(1,0), (3,1), (5,2), (7,3), (23,10)$; again see Figure 1. Similarly, the upper Christoffel path determines a sequence of integer points (following the *lower* boundary of its convex hull) beginning at $(0,1)$ and ending at (a,b). Let $\sigma(a,b)$ denote the sequence obtained from these two sets of integer points by deleting $(1,0)$ and $(0,1)$ and ordering the remaining points using the natural (lexicographic) ordering of $\mathbb{N}^2$. Let $\operatorname{ext}(a,b)$ denote the subsequence of $\sigma(a,b)$ consisting of the points that are *extreme points* of either of the convex hulls defined above. Recall that an **extreme point**

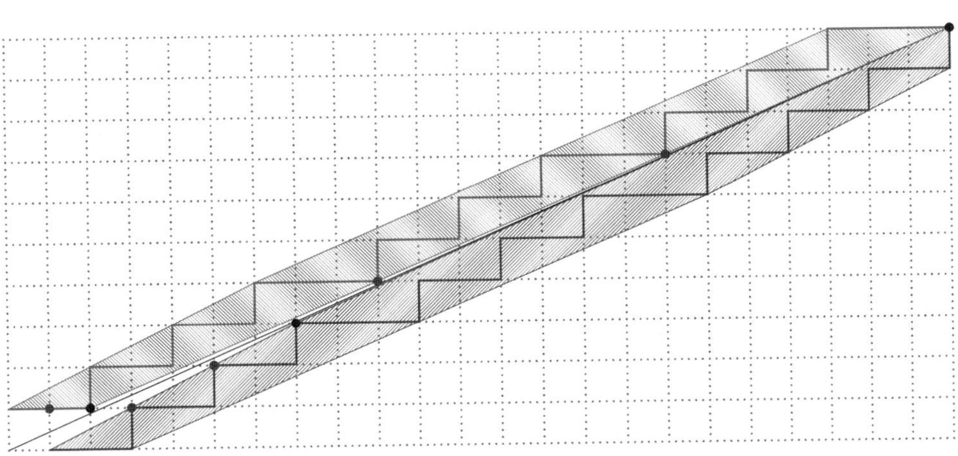

FIGURE 1. The convex hulls of the lower Christoffel path from $(0,1)$ to $(23,10)$ and the upper Christoffel path from $(1,0)$ to $(23,10)$.

of a convex set S in the plane is a point in S that does not lie in any open line segment joining two points of S.

Example. Let $b/a = 10/23$. Then the lower Christoffel path determines the sequence of points $(1,0)$, $(3,1)$, $(5,2)$, $(7,3)$, $(23,10)$; and the upper Christoffel path determines the sequence of points $(0,1)$, $(1,1)$, $(2,1)$, $(9,4)$, $(16,7)$, $(23,10)$. See Figure 1. The sequences $\sigma(23,10)$ and $\operatorname{ext}(23,10)$ are

$$\sigma(23,10)\colon (1,1),(2,1),(3,1),(5,2),(7,3),(9,4),(16,7),(23,10)$$

and

$$\operatorname{ext}(23,10)\colon (2,1),(7,3),(23,10).$$

Proposition 7.3. *Suppose a and b are positive integers and $a \perp b$. Let $(p_1,q_1),(p_2,q_2),\ldots,(p_s,q_s)$ denote the sequence $\operatorname{ext}(a,b)$. Then the ith continuant of b/a, for $1 \le i \le s$, is q_i/p_i.*

Example. Let $b/a = 10/23$. In the previous example, we found that $\operatorname{ext}(23,10)$ comprises the integer points $(2,1)$, $(7,3)$ and $(23,10)$. The proposition thus implies that $1/2$, $3/7$ and $10/23$ are the continuants of $10/23$, in agreement with the computation in (7.1).

We now describe how to obtain the sequences $\sigma(a,b)$ and $\operatorname{ext}(a,b)$ from the Christoffel tree. We begin with $\sigma(a,b)$.

Proposition 7.4. *Suppose a and b are positive integers and $a \perp b$. Let $(x,y) = (u_0,v_0),(u_1,v_1),\ldots,(u_r,v_r) = (u,v)$ denote the unique path in the Christoffel tree from (x,y) to (u,v) where uv is the Christoffel word of slope b/a. Then $\sigma(a,b)$ is the sequence of integer points*

$$(1,1),(|u_1v_1|_x,|u_1v_1|_y),(|u_2v_2|_x,|u_2v_2|_y),\ldots,(|uv|_x,|uv|_y) = (a,b).$$

Example. Continuing with the above example ($b/a = 10/23$), Figure 2 illustrates the unique path from the root (x,y) to the vertex labelled

$$(x^3yx^2yx^2y, x^3yx^2yx^2yx^3yx^2yx^2yx^2y)$$

and the corresponding sequence of integer points.

The sequence $\operatorname{ext}(a,b)$ can also be obtained from the Christoffel tree. Begin with the unique path $(x,y) = (u_0,v_0),(u_1,v_1),\ldots,(u_r,v_r) = (u,v)$ from (x,y) to (u,v), where uv is the Christoffel word of slope b/a. Let $(u_{i_1},v_{i_1}),(u_{i_2},v_{i_2}),\ldots,(u_{i_k},v_{i_k})$ denote the set of points immediately preceding a "bend" in the path, i.e., the points (u_j,v_j) for which the subpath from (u_j,v_j) to (u_{j+2},v_{j+2}) is one of the paths in Figure 3.

Proposition 7.5. *Suppose $a \perp b$ and let $(u_{i_1},v_{i_1}),(u_{i_2},v_{i_2}),\ldots,(u_{i_k},v_{i_k})$ denote the points constructed above. Then $\operatorname{ext}(a,b)$ is the sequence of integer points,*

$$(|u_{i_1}v_{i_1}|_x,|u_{i_1}v_{i_1}|_y),(|u_{i_2}v_{i_2}|_x,|u_{i_2}v_{i_2}|_y),\ldots,(|u_{i_k}v_{i_k}|_x,|u_{i_k}v_{i_k}|_y),(a,b).$$

Example. In Figure 2 we see that the vertices (x,xy) and (x^3yx^2y,x^2y) are the only vertices in the path that satisfy the conditions in Figure 3. Therefore, $\operatorname{ext}(23,10)$ is obtained from the three points (x,xy), (x^3yx^2y,x^2y) and $(x^3yx^2yx^2y, x^3yx^2yx^2yx^3yx^2yx^2yx^2y)$ by counting the number of occurrences of the letters x and y in these pairs of words: $\operatorname{ext}(23,10)$ is the sequence $(2,1),(7,3)$, $(23,10)$.

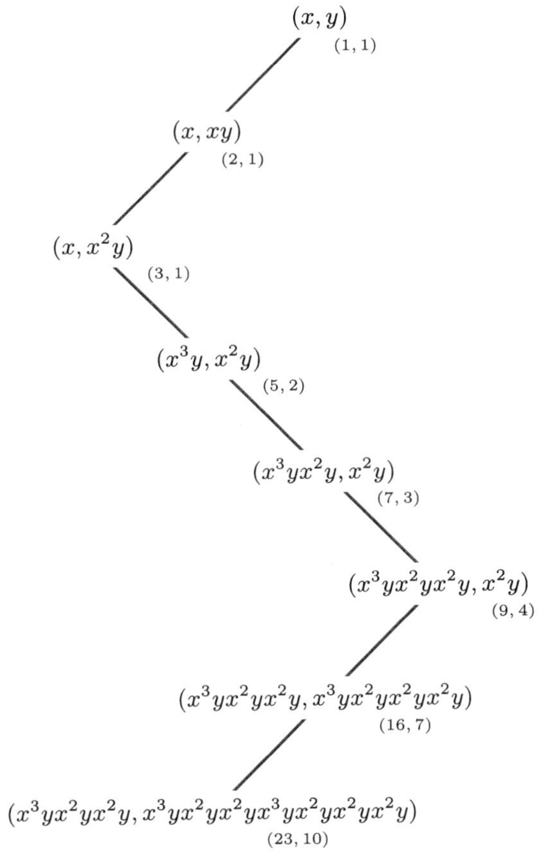

FIGURE 2. The sequences $\sigma(23, 10)$ and $\text{ext}(23, 10)$ can be obtained from the Christoffel tree by counting the number of occurrences of the letters x and y along the path from the root (x, y) to (u, v), where uv is the Christoffel word of slope of 10/23.

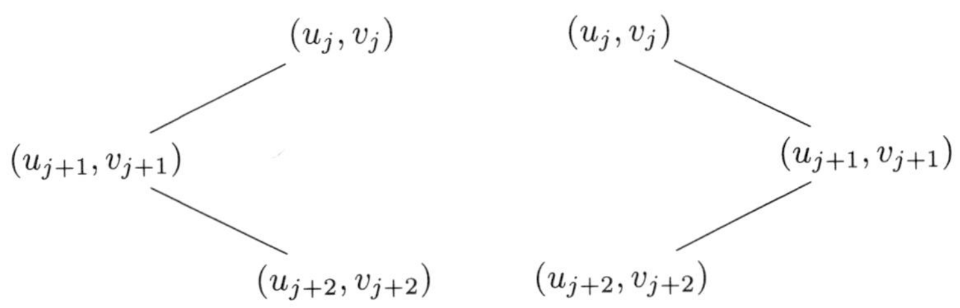

FIGURE 3. The points $(|u_jv_j|_x, |u_jv_j|_y)$ are in the sequence $\text{ext}(a, b)$.

Exercise 7.1. Find the continued fraction representations of the golden ratio $\phi = (1 + \sqrt{5})/2$ and its negated conjugate $-\phi^\vee = (\sqrt{5} - 1)/2$. Show that the nth continuant of ϕ is F_{n+1}/F_n, where F_n is the nth **Fibonacci number** defined recursively by $F_0 = F_1 = 1$ and $F_n = F_{n-1} + F_{n-2}$ for all $n \geq 2$.

Exercise 7.2. Rework Exercise 1.5 using Exercise 7.1 and Theorem 7.2.

3. The Stern – Brocot tree

The **mediant** of two fractions a/b and c/d is $(a+c)/(b+d)$. This operation gives rise to the Stern – Brocot tree, constructed recursively as follows. Let s_0 denote the sequence $0/1, 1/0$, where $1/0$ is viewed as a formal fraction. For $i > 0$, let s_i denote the sequence obtained from s_{i-1} by inserting between consecutive elements of the sequence their mediant. The first few iterations of this process yields the following sequences.

$$(s_0) \qquad \frac{0}{1}, \frac{1}{0}$$

$$(s_1) \qquad \frac{0}{1}, \frac{\mathbf{1}}{\mathbf{1}}, \frac{1}{0}$$

$$(s_2) \qquad \frac{0}{1}, \frac{\mathbf{1}}{\mathbf{2}}, \frac{1}{1}, \frac{\mathbf{2}}{\mathbf{1}}, \frac{1}{0}$$

$$(s_3) \qquad \frac{0}{1}, \frac{\mathbf{1}}{\mathbf{3}}, \frac{1}{2}, \frac{\mathbf{2}}{\mathbf{3}}, \frac{1}{1}, \frac{\mathbf{3}}{\mathbf{2}}, \frac{2}{1}, \frac{\mathbf{3}}{\mathbf{1}}, \frac{1}{0}$$

$$(s_4) \qquad \frac{0}{1}, \frac{\mathbf{1}}{\mathbf{4}}, \frac{1}{3}, \frac{\mathbf{2}}{\mathbf{5}}, \frac{1}{2}, \frac{\mathbf{3}}{\mathbf{5}}, \frac{2}{3}, \frac{\mathbf{3}}{\mathbf{4}}, \frac{1}{1}, \frac{\mathbf{4}}{\mathbf{3}}, \frac{3}{2}, \frac{\mathbf{5}}{\mathbf{3}}, \frac{2}{1}, \frac{\mathbf{5}}{\mathbf{2}}, \frac{3}{1}, \frac{\mathbf{4}}{\mathbf{1}}, \frac{1}{0}$$

The mediants constructed in the ith iteration of the above process, for $i > 0$, are the vertices in the ith level of the **Stern – Brocot tree**; there is an edge in the Stern – Brocot tree between a vertex a/b in level i and a vertex c/d in level $i-1$ if and only if a/b and c/d are consecutive elements of the sequence s_i. For example, there is an edge between $2/1$ and $3/2$ since $3/2$ and $2/1$ are consecutive elements of the sequence s_3:

$$\frac{0}{1}, \frac{1}{3}, \frac{1}{2}, \frac{2}{3}, \frac{1}{1}, \frac{\mathbf{3}}{\mathbf{2}}, \frac{\mathbf{2}}{\mathbf{1}}, \frac{3}{1}, \frac{1}{0}$$

Figure 4 shows the top 5 levels of the Stern – Brocot tree. Each fraction in the tree is of the form $(a+c)/(b+d)$, where a/b is the nearest ancestor above and to the right of $(a+c)/(b+d)$ and c/d is the nearest ancestor above and to the left of $(a+c)/(b+d)$.

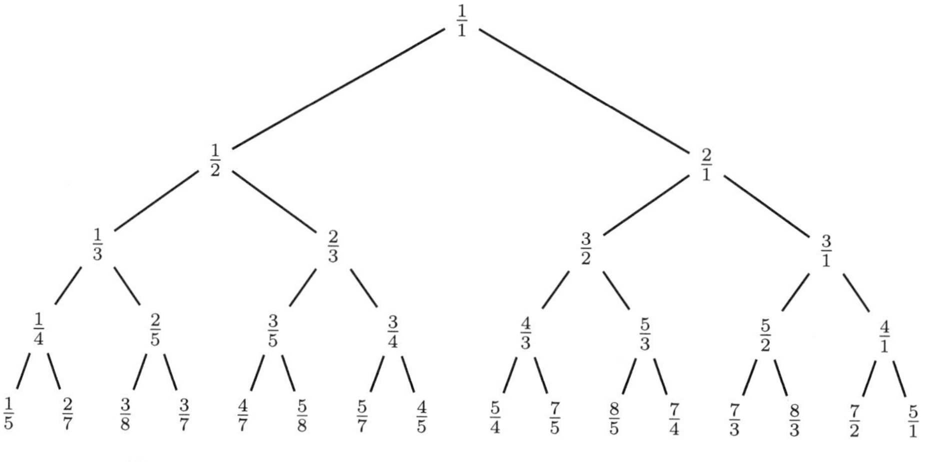

FIGURE 4. The first five levels of the Stern – Brocot tree.

Proposition 7.6. *The Christoffel tree is isomorphic to the Stern – Brocot tree via the map that associates to a vertex* (u, v) *of the Christoffel tree the fraction* $|uv|_y/|uv|_x$*. The inverse map associates to a fraction* b/a *the pair* (u, v)*, where* (u, v) *is the standard factorization of the Christoffel word of slope* b/a*.*

By Theorem 3.6, if $a \perp b$ and $a, b > 0$, then the standard factorization (u, v) of the Christoffel word of slope b/a appears exactly once in the Christoffel tree. Together with the above isomorphism, this implies the following classical result about the Stern – Brocot tree.

Corollary 7.7. *Every positive rational number appears in the Stern – Brocot tree exactly once.*

Moreover, Propositions 7.3 and 7.4 combine with Proposition 7.6 to give a method for determining the continuants of a real number from the Stern – Brocot tree. We leave the details to the interested reader.

The following exercises outline a proof of the fact that the Stern – Brocot tree contains each positive rational number exactly once without mention of Christoffel words. See [GKP1994, Chapter 4.5].

Exercise 7.3. Suppose a/b and c/d are connected by an edge of the Stern – Brocot tree. Then $(a + c) \perp (b + d)$. (*Hint*: Proceed by induction on the level of the fractions in the tree, and use *Bézout's Lemma* from Exercise 3.3.)

Exercise 7.4. If $a \perp b$ and $c \perp d$ and $a/b < c/d$, then $a/b < (a + c)/(b + d) < c/d$. Hence, each level of the Stern – Brocot tree preserves the natural order of the rational numbers.

Exercise 7.5. Suppose $a \perp b$. Then a/b is contained in the Stern – Brocot tree exactly once. (*Hint*: Use the previous exercise to show that a/b can occur at most once. To show that a/b occurs in the tree, begin with the sequence $0/1 < a/b < 1/0$ and take mediants. Argue that a/b is eventually the mediant of two fractions.)

CHAPTER 8

The Theory of Markoff Numbers

In 1879 and 1880, Andrey A. Markoff (later and better known as Markov) published two memoirs on the minima of indefinite binary quadratic forms [Mar1879, Mar1880]. He later used the results to answer a longstanding question of Bernoulli [Ber1771, Mar1881]. In this chapter, we reformulate some of the results from these memoirs in terms of Christoffel words.

1. Minima of binary quadratic forms

Let $f(x, y) = ax^2 + bxy + cy^2$ be a real binary quadratic form. The **discriminant** of f is $d(f) = b^2 - 4ac$, and the **minimum** of f is $m(f) = \inf |f(x, y)|$, where x and y range over all pairs of integers that are not both zero. Two binary forms f and g are **equivalent** if there exist $r, s, t, u \in \mathbb{R}$ such that $ru - st = \pm 1$ and $f(x, y) = g(rx + sy, tx + uy)$. If f and g are equivalent binary quadratic forms, then $d(f) = d(g)$ (Exercise 8.1).

Markoff's work was motivated by the following result of Alexander Korkine and Grigorii Zolotareff [KZ1873]. For any binary quadratic form f with $d(f) > 0$,

$$m(f) \leq \frac{\sqrt{d(f)}}{\sqrt{5}}, \tag{8.1}$$

with equality if and only if f is equivalent to a scalar multiple of

$$f_0(x, y) = x^2 - xy - y^2,$$

and if f is not equivalent to a scalar multiple of f_0, then $m(f) \leq \sqrt{d(f)}/\sqrt{8}$, with equality if and only if f is equivalent to a scalar multiple of

$$f_1(x, y) = x^2 - 2xy - y^2.$$

After learning of these results, Markoff set himself the task of finding the quantity that should replace $\sqrt{5}$ in (8.1) for forms that are not equivalent to scalar multiples of f_0 or f_1 [Mar1879]. He concluded that if f is such a form, then $m(f) \leq \sqrt{d(f)}/\sqrt{221/25}$, with equality if and only if f is equivalent to a scalar multiple of

$$f_2(x, y) = 5x^2 - 11xy - 5y^2.$$

Furthermore, he showed that this sequence of exceptions $(f_0, f_1, f_2, \ldots)$ and better approximations $(\sqrt{5}, \sqrt{8}, \sqrt{221/25}, \ldots)$ may be extended indefinitely.

Markoff's idea was to study a bi-infinite sequence of positive integers associated to a binary quadratic form g with positive discriminant. We briefly describe how he obtained this sequence. It is a well-known result (see, for example, Carl F. Gauss's *Disquisitiones Arithmeticae* [Gau1986] or [Dic1930, Chapter VII]) that any binary quadratic form g is equivalent to a *reduced* form $f(x, y) = ax^2 + bxy + cy^2$: the form f is said to be **reduced** if $f(x, 1) = ax^2 + bx + c$ has a positive root ξ

and a negative root η satisfying $|\eta| < 1 < \xi$. Writing $\xi = [a_0, a_1, a_2, \ldots]$ and $-\eta = [0, a_{-1}, a_{-2}, \ldots]$ for the continued fraction representations of ξ and $-\eta$, we obtain a bi-infinite sequence

$$A = (\ldots, a_{-2}, a_{-1}, a_0, a_1, a_2, \ldots).$$

Then $\sqrt{d(f)}/m(f)$ is equal to $\sup_{i\in\mathbb{Z}} \lambda_i(A)$, where

$$\lambda_i(A) = a_i + [0, a_{i+1}, a_{i+2} \ldots] + [0, a_{i-1}, a_{i-2} \ldots]. \tag{8.2}$$

Conversely, if $A = (\ldots, a_{-1}, a_0, a_1, \ldots)$ is a bi-infinite sequence of positive integers, then there exists a form f such that $\sqrt{d(f)}/m(f) = \sup_{i\in\mathbb{Z}} \lambda_i(A)$.

In the following we present results from [Mar1879, Mar1880] concerning the bi-infinite sequences of positive integers A such that $\sup_i \lambda_i(A) < 3$. As revealed in the work of Thomas W. Cusick and Mary E. Flahive [CF1989] and Christophe Reutenauer [Reu2006a, Reu2006b], Christoffel words make an unexpected appearance here. We recast Markoff's results in these terms. As a hint of what is to come, we rewrite the Markoff numbers $(\sqrt{5}, \sqrt{8}, \sqrt{221/25}, \ldots)$ in the form predicted by the general theory:

$$\sqrt{9 - \frac{4}{1^2}} < \sqrt{9 - \frac{4}{2^2}} < \sqrt{9 - \frac{4}{5^2}} < \cdots < 3.$$

The "squared" integers appearing above are explicitly computable and we find them below. Additional information on the Markoff theory may be found in [Fro1913], [Dic1930] and [Cas1957].

Exercise 8.1. Show that if f and g are equivalent binary quadratic forms, then they have the same discriminant.

2. Markoff numbers

We are interested in the bi-infinite sequences $A = (\ldots, a_{-1}, a_0, a_1, \ldots)$ where the a_i belong to the positive integers $\mathbb{P}$ (in what follows, we write $A \in \mathbb{P}^{\mathbb{Z}}$). Again, to such a sequence A we define positive real numbers $\boldsymbol{\lambda_i(A)} \in \mathbb{R}_{>0}$ by

$$\lambda_i(A) = a_i + [0, a_{i+1}, a_{i+2}, \ldots] + [0, a_{i-1}, a_{i-2}, \ldots],$$

where following Chapter 7, $[a_0, a_1, a_2, \ldots]$ denotes the limit as i goes to infinity of the ith continuant

$$[a_0, a_1, \ldots, a_i] := a_0 + \cfrac{1}{a_1 + \cfrac{1}{a_2 + \cfrac{1}{\cdots + \cfrac{1}{a_{i-1} + \cfrac{1}{a_i}}}}}.$$

Given A as above, denote the supremum of the $\lambda_i(A)$ by $\boldsymbol{M(A)}$:

$$M(A) = \sup_{i\in\mathbb{Z}}\{\lambda_i(A)\} \in \mathbb{R}_{>0} \cup \{\infty\}. \tag{8.3}$$

We will frequently view sequences $A \in \mathbb{P}^{\mathbb{Z}}$ or $B \in \mathbb{P}^{\mathbb{N}}$ as (infinite) words in what follows, e.g., $B = b_0 b_1 b_2 \cdots$. In such a case, we denote by $[B]$ the continued fraction $[b_0, b_1, b_2, \ldots]$.

Examples.

(1) Take $A = \cdots 111 \cdots$. Then $\lambda_i(A) = 1 + [0,1,\ldots] + [0,1,1\ldots]$ for each i, so we have $M(A) = (\sqrt{5}+1)/2 + (\sqrt{5}-1)/2 = \sqrt{5}$. See Exercise 7.1.

(2) Take $A = \cdots 222 \cdots$. Then $M(A)$ is computed to be $\sqrt{8}$.

Before stating our next two examples, we introduce some additional notation. If w is a finite word, let

$$^{\infty}w^{\infty} = \cdots www \cdots$$

denote the periodic bi-infinite word obtained by repeating w infinitely often in each direction. Also, let φ denote the morphism $\varphi\colon \{x,y\}^* \to \{1,2\}^*$ defined by

$$\varphi(x) = 11 \quad \text{and} \quad \varphi(y) = 22.$$

Finally, given any $A \in \mathbb{P}^{\mathbb{Z}}$, we define the reversal map $A \mapsto \tilde{A}$ by $a_i \mapsto a_{-i}$.

Examples.

(3) Take $A = {}^{\infty}\varphi(xy)^{\infty} = {}^{\infty}(1122)^{\infty}$. Since A is periodic, one need only compute $\lambda_i(A)$ for four consecutive values of i in order to determine $M(A)$. The result is computed to be $\sqrt{221/25}$.

(4) Let $\boldsymbol{w}$ denote the bi-infinite (nonperiodic) discretization of the line $\ell(x) = ((\sqrt{5}-1)/2)x$ (see Exercise 1.5). Taking $A = \varphi(\boldsymbol{w})$, one finds that $M(A)$ is approximately 3.2268.

(5) Given any $A \in \mathbb{P}^{\mathbb{Z}}$, we consider $\tilde{A}$. It is immediate that $\lambda_i(\tilde{A}) = \lambda_{-i}(A)$ for all $i \in \mathbb{Z}$ and $M(\tilde{A}) = M(A)$.

As observed above, $M(A)$ may be computed in a finite number of steps when A is periodic. This will always be the case if $M(A) < 3$, as the next theorem states. We need one more morphism in order to state it. Define $\eta\colon \{x,y\}^* \to \mathbb{N}^{2\times 2}$ to be the monoid morphism given by

$$\eta(x) = \begin{pmatrix} 1 & 1 \\ 1 & 0 \end{pmatrix}^2 = \begin{pmatrix} 2 & 1 \\ 1 & 1 \end{pmatrix} \quad \text{and} \quad \eta(y) = \begin{pmatrix} 2 & 1 \\ 1 & 0 \end{pmatrix}^2 = \begin{pmatrix} 5 & 2 \\ 2 & 1 \end{pmatrix}.$$

Theorem 8.4 (Markoff [Mar1879, Mar1880]**).** *Let $A \in \mathbb{P}^{\mathbb{Z}}$ be a bi-infinite sequence of positive integers. Then $M(A) < 3$ if and only if there exists a Christoffel word w such that $A = {}^{\infty}\varphi(w)^{\infty}$. In this case, $M(A) = \sqrt{9 - 4/c^2}$, where c is the $(1,2)$-coordinate of the matrix $\eta(w)$.*

Remark. The numbers c obtained in the theorem are the so-called **Markoff numbers**. It happens that c is also equal to $\frac{1}{3}\operatorname{trace}(\eta(w))$; the interested reader may try to prove this now or wait for the hint in Lemma 8.7.

Examples.

(1) Take $w = x$ (i.e., $A = \cdots 111 \cdots$). Then

$$\eta(w) = \begin{pmatrix} 2 & 1 \\ 1 & 1 \end{pmatrix},$$

$c = 1$ and $\sqrt{9 - 4/1^2} = \sqrt{5}$ agrees with the value for $M(A)$ computed in the preceding example.

(2) Taking $w = y$, we find $c = 2$ and $M(\cdots 222 \cdots) = \sqrt{9 - 4/2^2} = \sqrt{8}$, as computed in the preceding example.

(3) Suppose w is the Christoffel word xy. Then $A = {}^{\infty}(1122)^{\infty}$ and

$$\eta(w) = \begin{pmatrix} 2 & 1 \\ 1 & 1 \end{pmatrix}\begin{pmatrix} 5 & 2 \\ 2 & 1 \end{pmatrix} = \begin{pmatrix} 12 & 5 \\ 7 & 3 \end{pmatrix},$$

giving $c = 5$ and $M(A) = \sqrt{221/25}$.

Before proceeding to the proof of Theorem 8.4 (Section 4), we raise a long-standing open question. Implicit in the theorem is a mapping $w \mapsto c(w)$ of Christoffel words to Markoff numbers. As $c(w)$ is somewhat correlated to $|w|$, it is plausible (indeed, provable) that the image of c in $\mathbb{P}$ is not surjective. The complimentary question remains open.

Open Question (Frobenius [Fro1913]**).** *Is the mapping $w \mapsto c(w)$ injective?*

Exercise 8.2. The map from words w to Markoff numbers $c(w) = \big(\eta(w)\big)_{12}$ may be injective on Christoffel words, but it is not injective on all of $\{x, y\}^*$. (*Hint*: Use the word $w = xxyy$.)

Exercise 8.3. Given fixed morphisms $\alpha, \beta \in \operatorname{End}(\{x, y\}^*)$, define a *representation* $\rho\colon \{x, y\}^* \to \operatorname{End}(\{x, y\}^*)$ of the monoid $\{x, y\}^*$ by sending x to α, y to β and concatenation to composition. For example, ρ_{xyy} is the morphism

$$w \mapsto (\alpha \circ \beta \circ \beta)(w).$$

Show that there are Christoffel morphisms α and β so that $\big|\rho_w(y)\big|_x$ equals the Markoff number $c(w)$ for all words $w \in \{x, y\}^*$. (*Hint*: Try to mimic the action of the map η above; as a further hint, α and β both belong to $\{\mathbf{G}, \widetilde{\mathbf{D}}\}^*$.)

3. Markoff's condition

Lemma 8.5. *Suppose β and γ are two real numbers with continued fraction representations $[b_1, b_2, b_3, \ldots]$ and $[c_1, c_2, c_3, \ldots]$, respectively. Determining whether $\beta \leq \gamma$ amounts to comparing finite prefixes $b_1 b_2 \cdots b_i$ and $c_1 c_2 \cdots c_i$ with the following rules:*

$$\beta < \gamma \iff \begin{cases} b_1 < c_1, \\ \text{or } b_1 = c_1 \text{ and } b_2 > c_2, \\ \text{or } b_1 = c_1 \text{ and } b_2 = c_2 \text{ and } b_3 < c_3, \\ \text{and so on.} \end{cases}$$

PROOF. Exercise 8.4. □

Given two (possibly infinite) words $B = b_1 b_2 \cdots$ and $C = c_1 c_2 \cdots$, we say that B precedes C in the **alternating lexicographic order** if the prefixes $b_1 b_2 \cdots b_i$ and $c_1 c_2 \cdots c_i$ satisfy the conditions of the lemma for some $i > 0$.

Given a word w over an alphabet X, a letter $x \in X$ and any $p \in \mathbb{P}$, the word x^p is called a **block** of w if there is a factorization $w = ux^pv$ such that $u \in X^*$ does not end by x and $v \in X^*$ does not begin by x.

Lemma 8.6 (Markoff). *Given $A \in \mathbb{P}^{\mathbb{Z}}$, if $M(A) \leq 3$ then $a_i \in \{1, 2\}$ for all $i \in \mathbb{Z}$ and the blocks of 1s and 2s are of even length.*

Note that the converse is false, as evidenced by the "Fibonacci line" constructed in Example 4.

PROOF. First note that $\lambda_i(A)$ is equal to a_i plus a positive number. This forces $a_i < 3$, i.e., $a_i \in \{1, 2\}$ for all $i \in \mathbb{Z}$. We show that 12^n1 and 21^n2 are not factors of A for odd numbers n by induction on n. The case $n = 1$ is a simple calculation:

If **121** is a factor of A, choose i to be the position of this 2. Then

$$\lambda_i(A) = \mathbf{2} + \cfrac{1}{\mathbf{1} + \cfrac{1}{*_1}} + \cfrac{1}{1 + \cfrac{1}{*_2}}.$$

Since each $*_i > 1$, we have $1 + 1/*_i < 1 + 1/1$, or $1/(1 + 1/*_i) > 1/(1 + 1)$. But then

$$\lambda_i(A) > 2 + \tfrac{1}{2} + \tfrac{1}{2} = 3,$$

contradicting the hypothesis $M(A) \leq 3$. If **212** is a factor of A, choose i to be the position of the second 2 and write

$$\lambda_i(A) = \mathbf{2} + \frac{1}{*_1} + \cfrac{1}{1 + \cfrac{1}{2 + \cfrac{1}{*_2}}}.$$

Since $*_1 < 3$, the second summand is bounded below by $\frac{1}{3}$. Turning to the final summand, we use the inequality $1/*_2 > 0$. This yields

$$\cfrac{1}{1 + \cfrac{1}{2 + \cfrac{1}{*_2}}} > \cfrac{1}{1 + \cfrac{1}{2 + 0}} = \frac{2}{3},$$

or $\lambda_i(A) > 2 + \frac{1}{3} + \frac{2}{3} = 3$, again contradicting the hypothesis $M(A) \leq 3$.

To rule out factors 21^n2 and 12^n1 for odd $n > 1$, we observe the following useful fact (see Exercise 8.5):

($\star$) *If $A \in \mathbb{P}^{\mathbb{Z}}$ has a factorization $\widetilde{B}\,2211\,C$ with B and C right-infinite words, then $M(A) \leq 3$ implies $[B] \leq [C]$.*

From this fact we deduce that A cannot be factored as $\cdots 2(2211)1 \cdots$, because the integral part of $[B] = [2, \ldots]$ would be greater than that of $[C] = [1, \ldots]$ (a contradiction after Lemma 8.5).

Case 1: *there is a factor* 21^n2 *in* A *with* $n > 1$ *and* n *odd.*

We analyze the slightly longer factor $1^r2^s1^n2$ with $s \geq 1$ and r maximal (possibly infinite). We know that 212, 121 and 222111 are not factors of A. These exclude, respectively, the possibilities $r = 1$, $s = 1$ or $s \geq 3$. We are left with $s = 2$ and the two possibilities: (i) $21^r(2211)1^{n-2}2$, with $r < n - 1$, is a factor of A; or (ii) $1^r(2211)1^{n-2}2$, with $r \geq n - 1$, is a factor of A. We may apply induction in the first possibility to further assume that n is even and less than $n - 2$.

Comparing $[B]$ to $[C]$ using Lemma 8.5, both possibilities yield a contradiction according to ($\star$):

$$
\text{(i)}\quad
\begin{array}{cccccc}
 & \multicolumn{4}{c}{\overbrace{\hphantom{1\quad 1\quad \cdots\quad 1}}^{r,\ \text{even}}} & \\
B: & 1 & 1 & \cdots & 1 & 2 \\
 & /\!\!\wedge & \backslash\!\!\vee & \cdots & \backslash\!\!\vee & \not{/\!\!\wedge} \\
C: & 1 & 1 & \cdots & 1 & 1
\end{array}
\qquad\qquad
\text{(ii)}\quad
\begin{array}{cccccc}
B: & 1 & 1 & \cdots & 1 & 1 \\
 & /\!\!\wedge & \backslash\!\!\vee & \cdots & /\!\!\wedge & \not{\backslash\!\!\vee} \\
C: & 1 & 1 & \cdots & 1 & 2 \\
 & \multicolumn{4}{c}{\underbrace{\hphantom{1\quad 1\quad \cdots\quad 1}}_{n-2,\ \text{odd}}} &
\end{array}
$$

Case 2: *there is a factor* 12^n1 *in* A *with* $n > 1$ *and* n *odd.*

The analysis is similar. Assume $12^n1^s2^r$ is a factor of A (with $s \geq 1$ and r maximal). As above, one easily reduces to the case $s = 2$. The remaining possibilities, $r < n - 2$ and even or $r \geq n - 1$, are again handled using ($\star$) and Lemma 8.5. □

Lemma 8.7. *Fix a* 2×2 *symmetric matrix* M *and set*

$$N = \begin{pmatrix} 5 & 2 \\ 2 & 1 \end{pmatrix} M \begin{pmatrix} 2 & 1 \\ 1 & 1 \end{pmatrix}.$$

Then $N_{21} = \frac{1}{3} \operatorname{trace} N$.

PROOF. Exercise 8.6. □

Corollary 8.8. *If* $w = yux$ *is an upper Christoffel word, then* $\eta(u)$ *is symmetric and* $\big(\eta(w)\big)_{21} = \frac{1}{3} \operatorname{trace}\big(\eta(w)\big)$.

Markoff introduced the following condition to exploit this property in his proof of Theorem 8.4.

Definition 8.9. Suppose $\boldsymbol{s} \in \{x, y\}^{\mathbb{Z}}$. We say $\boldsymbol{s}$ satisfies the **Markoff condition** if for each factorization $\boldsymbol{s} = \tilde{\boldsymbol{u}}ab\boldsymbol{v}$ with $\{a, b\} = \{x, y\}$, one has either $\boldsymbol{u} = \boldsymbol{v}$, or $\boldsymbol{u} = mb\boldsymbol{u}'$ and $\boldsymbol{v} = ma\boldsymbol{v}'$ for some (possibly empty) finite word m and right-infinite words $\boldsymbol{u}', \boldsymbol{v}'$.

Examples.

(1) Take $\boldsymbol{s} = {}^{\infty}(xxy)^{\infty}$. Then the factorization

$$\tilde{\boldsymbol{u}} \cdot ab \cdot \boldsymbol{v} = {}^{\infty}(xxy)xx \cdot yx \cdot xy(xxy)^{\infty}$$

yields $\boldsymbol{u} = m \cdot b \cdot \boldsymbol{u}' = x \cdot x \cdot (yxx)^{\infty}$ and $\boldsymbol{v} = m \cdot a \cdot \boldsymbol{v}' = x \cdot y \cdot (xxy)^{\infty}$. On the other hand, the factorization

$$\tilde{\boldsymbol{u}} \cdot ab \cdot \boldsymbol{v} = {}^{\infty}(xxy)xxyx \cdot xy \cdot xxy(xxy)^{\infty}$$

yields $\boldsymbol{u} = m \cdot b \cdot \boldsymbol{u}' = x \cdot y \cdot (xxy)^{\infty}$ and $\boldsymbol{v} = m \cdot a \cdot \boldsymbol{v}' = x \cdot x \cdot (yxx)^{\infty}$.

(2) Taking $\boldsymbol{s} = {}^{\infty}(xxyxy)^{\infty}$, the pattern ab appears in two distinct ways as "xy" and two distinct ways as "yx." We check one of the four possibilities and leave the rest for the reader. The factorization

$$\boldsymbol{s} = {}^{\infty}(xxyxy)\, x\, xyx \cdot yx \cdot xyx\, y\, (xxyxy)^{\infty}$$

yields $m = xyx$, $\boldsymbol{u} = xyx \cdot x \cdot (yxyxx)^{\infty}$ and $\boldsymbol{v} = xyx \cdot y \cdot (xxyxy)^{\infty}$.

(3) The bi-infinite word $\boldsymbol{s} = {}^{\infty}(xxyy)^{\infty}$ does not satisfy the Markoff condition (both $ab = xy$ and $ab = yx$ fail to give factorizations for $\boldsymbol{s}$ satisfying the criteria of the definition).

Remarks.

(1) Reutenauer has shown that a bi-infinite word $\boldsymbol{s}$ satisfies the Markoff condition if and only if it is balanced$_1$ ([Reu2006b, Theorem 3.1]).

(2) Note that in the first two examples above, m is a palindrome. This is always the case; a proof is outlined in Exercise 8.8 (see also [GLS2008]).

Exercise 8.4. Prove Lemma 8.5.

Exercise 8.5. Two useful properties of λ_i:

(a) Suppose $A = \widetilde{B}\,2211\,C$ is a factorization with B and C right-infinite words. If i is the location of the second 2 after $\widetilde{B}$, show that $\lambda_i(A) \le 3$ if and only if $[B] \le [C]$ (with $\lambda_i(A) = 3$ if and only if $[B] = [C]$).

(b) Suppose $A = \widetilde{B}\,1122\,C$ is a factorization with B and C right-infinite words. If i is the location of the first 2 after $\widetilde{B}$, show that $\lambda_i(A) \le 3$ if and only if $[B] \ge [C]$ (with $\lambda_i(A) = 3$ if and only if $[B] = [C]$).

Exercise 8.6. Prove Lemma 8.7 and deduce Corollary 8.8. (*Hint*: If $w = yux$ is an upper Christoffel word, then u is a palindrome.)

4. Proof of Markoff's theorem

After the preceding lemmas, it will be easier to prove the following equivalent statement of Theorem 8.4.

Theorem 8.10. *A bi-infinite sequence $A \in \mathbb{P}^{\mathbb{Z}}$ satisfies $M(A) < 3$ if and only if there exists an upper Christoffel word w such that $A = {}^{\infty}\varphi(w)^{\infty}$. In this case, $M(A) = \sqrt{9 - 4/c^2}$, where $c = \big(\eta(w)\big)_{21} = \frac{1}{3}\operatorname{trace}\big(\eta(w)\big)$.*

Suppose $A \in \mathbb{P}^{\mathbb{Z}}$ satisfies $M(A) < 3$. We have seen that A belongs to $\{1,2\}^{\mathbb{Z}}$ and moreover its blocks of 1s and 2s have even lengths. We may thus write $A = \varphi(\boldsymbol{s})$ for some $\boldsymbol{s} \in \{x,y\}^{\mathbb{Z}}$. We begin by showing that $\boldsymbol{s}$ satisfies the Markoff condition.

Given a factorization $\boldsymbol{s} = (\tilde{\boldsymbol{u}}, yx, \boldsymbol{v})$, we may write $A = \widetilde{B}\,2211\,C$ for some $B = \varphi(\boldsymbol{u})$ and $C = \varphi(\boldsymbol{v})$ in $\{11,22\}^{\mathbb{P}}$. From the assumption $M(A) \le 3$ and $(\star)$ we have $[B] \le [C]$, or $B \le C$ in the alternating lexicographic order. Equivalently, since $B, C \in \{11,22\}^{\mathbb{P}}$, we have $\boldsymbol{u} \le \boldsymbol{v}$ in the natural lexicographic order on $\{x,y\}^*$. If $\boldsymbol{u} = \boldsymbol{v}$, then $[B] = [C]$ and $M(A) = 3$ (see Exercise 8.5), which was excluded in the hypotheses of the theorem. Thus $\boldsymbol{u} < \boldsymbol{v}$. Letting $m = u_1 u_2 \cdots u_r$ be the longest common prefix of $\boldsymbol{u}$ and $\boldsymbol{v}$, we have $u_{r+1} = x$ and $v_{r+1} = y$ (since $\boldsymbol{u} < \boldsymbol{v}$). Analysis of the factorization $\boldsymbol{s} = (\tilde{\boldsymbol{u}}, xy, \boldsymbol{v})$ is similar (see Exercise 8.5).

We conclude that $\boldsymbol{s}$ satisfies the Markoff condition, but in fact more is true. Namely, the m's occuring in instances of the Markoff condition have bounded length $N = N(\boldsymbol{s})$ (depending only on $\boldsymbol{s}$). Otherwise, we may find an infinite sequence of factors $x\tilde{m}_n\,yx\,m_n y$ of $\boldsymbol{s}$ (or $\tilde{\boldsymbol{s}}$) satisfying m_n is a proper prefix of m_{n+1} for all n. One uses these factors and the ideas in Exercise 8.5 to show that $M(A) = \sup\lambda_i(A) = 3$, contradicting the hypotheses of the theorem.

Lemma 8.11 (Reutenauer [Reu2006b, Lemma 3.1]**).** *If $\boldsymbol{s} \in \{x,y\}^{\mathbb{Z}}$ satisfies the Markoff condition, then xx and yy are not simultaneously factors of $\boldsymbol{s}$.*

Proof. Suppose xx and yy are both factors of $\boldsymbol{s}$. Then $\boldsymbol{s} = \boldsymbol{u}'xxwyy\boldsymbol{v}'$ or $\boldsymbol{u}'yywxx\boldsymbol{v}'$ for some finite word w and some infinite words $\boldsymbol{u}'$ and $\boldsymbol{v}'$. Pick w to be of minimal length among all such finite words. Suppose $\boldsymbol{s} = \boldsymbol{u}'xxwyy\boldsymbol{v}'$; the other case is dealt with in a similar manner. By the minimality of w, we have that $w = (yx)^p$ for some $p \in \mathbb{N}$. If $p = 0$, then $\boldsymbol{s} = \boldsymbol{u}'xxyy\boldsymbol{v}'$, which contradicts the hypothesis that $\boldsymbol{s}$ satisfies the Markoff condition. So suppose $p \ge 1$. Let $\boldsymbol{u} = x\tilde{\boldsymbol{u}}'$

and $\boldsymbol{v} = (xy)^p y\boldsymbol{v}'$. Then $\boldsymbol{s} = \tilde{\boldsymbol{u}}xy\boldsymbol{v}$. Since $\boldsymbol{s}$ satisfies the Markoff condition, there are two cases.

Case 1: *the words* $\boldsymbol{u}$ *and* $\boldsymbol{v}$ *are equal.*

Then $\boldsymbol{s} = \tilde{\boldsymbol{v}}xy\boldsymbol{v} = \tilde{\boldsymbol{v}}'y(yx)^p xy\boldsymbol{v} = \tilde{\boldsymbol{v}}'yy(xy)^{p-1}xxy\boldsymbol{v}$, contradicting the minimality of w.

Case 2: *there exist right-infinite words* $\boldsymbol{u}''$ *and* $\boldsymbol{v}''$ *and a finite word* m *such that* $\boldsymbol{u} = my\boldsymbol{u}''$ *and* $\boldsymbol{v} = mx\boldsymbol{v}''$.

Since $\boldsymbol{u} = x\tilde{\boldsymbol{u}}'$ and $\boldsymbol{v} = (xy)^p y\boldsymbol{v}'$, we have that m is nonempty. If $|m| \geq 2p+1$, then m begins with $(xy)^p y$. So we have $\boldsymbol{s} = \tilde{\boldsymbol{u}}xy\boldsymbol{v} = \cdots y(yx)^p xy\boldsymbol{v} = \cdots yy(xy)^{p-1}xxxy\boldsymbol{v}$, contradicting the minimality of w. If $|m| < 2p+1$, then m is a prefix of $(xy)^p y$. Therefore, $m = (xy)^i$ for some $1 < i < p-1$. This implies $\boldsymbol{s} = \tilde{\boldsymbol{u}}xy\boldsymbol{v} = \tilde{\boldsymbol{u}}''y(yx)^i xy\boldsymbol{v} = \tilde{\boldsymbol{u}}''yy(xy)^{i-1}xxy\boldsymbol{v}$, again contradicting the minimality of w. □

Next, we lift sequences $\boldsymbol{s}$ satisfying the Markoff condition via the morphisms $\mathbf{G} = (x, xy)$ and $\widetilde{\mathbf{D}} = (xy, y)$. We claim there exists a sequence $\boldsymbol{s}' \in \{x,y\}^{\mathbb{Z}}$ such that $\boldsymbol{s} = \mathbf{G}(\boldsymbol{s}')$ or $\boldsymbol{s} = \widetilde{\mathbf{D}}(\boldsymbol{s}')$ (apply $\mathbf{G}^{-1}$ if yy is not a factor of $\boldsymbol{s}$ and $\widetilde{\mathbf{D}}^{-1}$ otherwise). It is straightforward (though tedious) to verify that $\boldsymbol{s}'$ also satisfies the Markoff condition (Exercise 8.8) and moreover that the bounds on the $|m|$'s satisfy $N(\boldsymbol{s}') < N(\boldsymbol{s})$, cf. [Reu2006b, Section 3]. An induction on $N(\boldsymbol{s})$ allows us to write $\boldsymbol{s}'$ as ${}^{\infty}(w')^{\infty}$ for some upper Christoffel word w' (note that $\boldsymbol{s} = {}^{\infty}(yx)^{\infty}$ when $N(\boldsymbol{s}) = 0$). Thus $A = {}^{\infty}\varphi(w)^{\infty}$ for some upper Christoffel word w, as desired.

To prove the converse, we write $A = {}^{\infty}\varphi(w)^{\infty}$ for some upper Christoffel word w and compute $M(A)$ explicitly.

Example. Suppose $A = {}^{\infty}\varphi(yxyxx)^{\infty}$. In Figure 1, we compute the first few $\lambda_i(A)$.

Returning to the proof, there are evidently only $2|w|$ distinct $\lambda_i(A)$ to compute, but we can do better. Since $\widetilde{w}$ is a conjugate of w (Proposition 4.2), we have $\tilde{A} = {}^{\infty}\varphi(\widetilde{w})^{\infty} = {}^{\infty}\varphi(w)^{\infty} = A$. Consequently, we need only consider those i corresponding to 1s and 2s in odd positions within their corresponding blocks in $\varphi(w)$.

We introduce some notation to make things simpler. Index the sequence $A = (a_n)_{n\in\mathbb{Z}}$ by setting $n = 1$ to be the location of the start of some copy of $\varphi(w)$ (in particular, $a_1 = 2$, see Figure 1). We have seen that it is enough to compute λ_i for i corresponding to odd integers. We claim it is enough to compute λ_1. Indeed, we

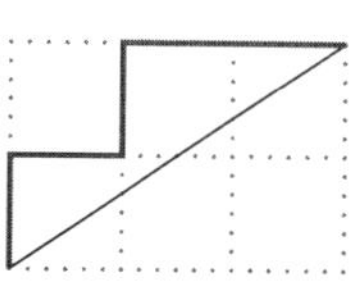

$A = \quad \cdots 1\,2\,2\,1\,1\,2 \cdots$
$\cdots\ 0\ 1\ 2\ 3\ 4\ 5\ \cdots$
⟶

i	λ_i
1	$\frac{2}{97}\sqrt{21170} \approx 2.999982286$
2	2.969370222
3	2.124075049
4	2.124075049
5	2.969370224
⋮	

FIGURE 1. The upper Christoffel word $yxyxx$ and some of the 10 possible $\lambda_i(A)$ values. They were computed using the technique indicated in the proof of Lemma 8.14.

now show that $\lambda_1 > \lambda_j$ when $j \not\equiv 1 \bmod 2|w|$ corresponds to any other odd integer, i.e., $M(A) = \lambda_1(A)$.

Our proof uses the fact that upper and lower Christoffel words are the least and greatest lexicographically among the conjugates of either (see Exercise 6.3). We compare w (an upper Christoffel word), $\widetilde{w}$ (the corresponding lower Christoffel word) and some other conjugate u of w. We have $w > u$ (lexicographically), which implies $w^\infty > u^\infty$ and $\varphi(w)^\infty > \varphi(u)^\infty$ (in the alternating lexicographic order). By Lemma 8.5, this in turn implies that $[\varphi(w)^\infty] > [\varphi(u)^\infty]$ (as real numbers). Similarly, $[{}^\infty\varphi(\widetilde{w})] < [{}^\infty\varphi(u)]$. In terms of our sequence, we have

$$\begin{aligned}[][a_1, a_2, \ldots] &> [a_j, a_{j+1}, \ldots],\\ [a_0, a_{-1}, \ldots] &< [a_{j-1}, a_{j-2}, \ldots],\end{aligned}$$

or equivalently

$$\begin{aligned}[][a_1, a_2, \ldots] &> [a_j, a_{j+1}, \ldots],\\ [0, a_0, a_{-1}, \ldots] &> [0, a_{j-1}, a_{j-2}, \ldots].\end{aligned}$$

Thus $\lambda_1(A) > \lambda_j(A)$, proving our assertion that $M(A) = \lambda_1(A)$.

To understand the Markoff numbers $c = 1, 2, 5, \ldots$ we need three classical facts about the continued fraction representations of *quadratic numbers:* $\alpha \in \mathbb{R}$ is a **quadratic number** if it is a solution to a monic, quadratic polynomial over $\mathbb{Q}$. In what follows, we use the notation $\overline{c_1, c_2, \ldots, c_p}$ to represent the periodic, right-infinite sequence $c_1, c_2, \ldots, c_p, c_1, c_2, \ldots, c_p, \ldots$. Also, given a quadratic number $\alpha = a \pm \sqrt{b}/c$, we write $\alpha^\vee$ for its **conjugate root**, i.e., $\alpha^\vee = a \mp \sqrt{b}/c$.

Theorem 8.12 (Lagrange). *A number $\alpha \in \mathbb{R}$ is quadratic if and only if its continued fraction representation is ultimately periodic. That is,*

$$\alpha = [c_0, c_1, \ldots, c_{r-1}, \overline{c_r, c_{r+1}, \ldots, c_{r+p-1}}].$$

Theorem 8.13 (Galois). *If a real number $\alpha = [c_0, c_1, c_2, \ldots]$ is quadratic, then the sequence (c_n) is purely periodic with period p if and only if $\alpha > 1$ and $-\alpha^\vee \in (0,1)$. In this case, $\alpha = [\overline{c_0, c_1, \ldots, c_{p-1}}]$ and $-1/\alpha^\vee = [\overline{c_{p-1}, \ldots, c_1, c_0}]$.*

Lemma 8.14. *If $\alpha = [\overline{c_0, c_1, \ldots, c_{p-1}}]$, then*

$$\alpha = \frac{a\alpha + b}{c\alpha + d}, \quad \textit{where} \begin{pmatrix} a & b \\ c & d \end{pmatrix} = \begin{pmatrix} c_0 & 1 \\ 1 & 0 \end{pmatrix} \cdot \cdots \cdot \begin{pmatrix} c_{p-1} & 1 \\ 1 & 0 \end{pmatrix}.$$

Sketch of Proof. We illustrate the key idea for $p = 2$ and leave the proof as an exercise. Suppose $\alpha = [\overline{a, b}] = [a, b, a, b, a, b, \cdots]$. Then

$$\alpha = a + \cfrac{1}{b + \cfrac{1}{a + \cfrac{1}{\ddots}}} = a + \cfrac{1}{b + \cfrac{1}{\alpha}}.$$

That is, $\alpha = a + \alpha/(b\alpha + 1) = ((ab+1)\alpha + a)/(b\alpha + 1)$. Compare with

$$\begin{pmatrix} a & 1 \\ 1 & 0 \end{pmatrix} \begin{pmatrix} b & 1 \\ 1 & 0 \end{pmatrix}.$$

□

From the above facts we deduce that $M(A) = \lambda_1(A) = a_1 + [0, a_2, \ldots] + [0, a_0, a_{-1}, \ldots] = \alpha - \alpha^\vee$, where $\alpha = [\varphi(w)^\infty]$. Moreover, the matrix in Lemma 8.14 for our α is precisely $\eta(w)$. Now let's find a, b, c, d explicitly. From

$$c\alpha^2 + d\alpha = a\alpha + b,$$

we deduce

$$\begin{aligned}\alpha - \alpha^\vee &= \frac{1}{c}\sqrt{(d-a)^2 + 4bc}\\ &= \frac{1}{c}\sqrt{(d+a)^2 - 4}\end{aligned}$$

(this last step because $\det \eta(w) = 1$, by definition of $\eta(x)$ and $\eta(y)$, so $bc = ad - 1$). Finally, we know from Lemma 8.7 that $3c = a + d$, i.e.,

$$M(A) = \lambda_1(A) = \frac{1}{c}\sqrt{9c^2 - 4} = \sqrt{9 - \frac{4}{c^2}},$$

concluding the proof of Markoff's theorem. □

In closing, we mention that the values $M(A)$ computed above are part of a larger story. The set of values $M(A)$ for all bi-infinite sequences A, not only those obeying Markoff's restrictions, is called the **Markoff spectrum**. Further results on this spectrum may be found in [CF1989].

Exercise 8.7. Prove Lemma 8.14.

The following exercise shows that any m from Definition 8.9 is a palindrome and that xmy is a lower Christoffel word or yux is an upper Christoffel word. (The proof in [GLS2008] uses more balanced$_1$ results than we have set down in Chapter 6.)

Exercise 8.8 ([GLS2008]). Let $\boldsymbol{s}$ be an infinite word in the letters x and y satisfying the Markoff condition.

(a) Prove that if yy is not a factor of $\boldsymbol{s}$, then $\mathbf{G}^{-1}(\boldsymbol{s})$ satisfies the Markoff condition; likewise for $\widetilde{\mathbf{D}}^{-1}(\boldsymbol{s})$ when xx is not a factor of $\boldsymbol{s}$. (See [Reu2006b].)

(b) Prove that if two (maximal) blocks x^a and x^b are factors of $\boldsymbol{s}$, then $|a-b| \leq 1$; likewise for blocks consisting of the letter y.

(c) Consider a factor $y\tilde{m}xymx$ of $\boldsymbol{s}$. If m starts with an x, conclude that m takes the form $x^a yx^b yx^c y \cdots yx^a$ (including the possibility $m = x^a$). Moreover, x^a is the smallest block of x that is a factor of $\boldsymbol{s}$.

(d) If u is a palindrome, then $(\mathbf{G}(u))^a x^a$ is a palindrome for all $a \geq 1$.

(e) Prove that m is a palindrome. (*Hint*: Proceed by induction on the number of x-blocks in m. Consider the preimage of s under $\mathbf{G}$ or $\widetilde{\mathbf{D}}$.)

Part II

Repetitions in Words

The goal of Part II is to present an introduction to some of the recent research on the combinatorics on words that deals with repetitions in words. The discipline originated in a series of papers by the Norwegian mathematician Axel Thue (1863–1922). Chief among them, we count [Thu1906] and [Thu1912]. Thue's work, recollected in the volume [Thu1977], has inspired several directions of modern research, so we have chosen to use his results as a point of departure for the results presented here.

CHAPTER 1

The Thue – Morse Word

This chapter introduces the Thue – Morse word and presents several equivalent characterizations. We end with a novel application of the Thue – Morse word to construct magic squares.

1. The Thue – Morse word

Recall that a **binary** word is a word over the alphabet $\{0, 1\}$.

Definition 1.1. The **Thue – Morse word** $\boldsymbol{t} = t_0t_1t_2\cdots$ is the binary word $\boldsymbol{t}\colon \mathbb{N} \to \{0,1\}$ defined recursively by: $t_0 = 0$; and for $n \geq 0$, $t_{2n} = t_n$ and $t_{2n+1} = \bar{t}_n$, where $\bar{a} = 1 - a$ for $a \in \{0,1\}$. (See Figure 1.)

Example. Here are the first forty letters of the Thue – Morse word,

$$\boldsymbol{t} = 0110100110010110100101100110100110010110\cdots$$

Our first characterization of the Thue – Morse word is in terms of binary expansions of nonnegative integers. For every $n \in \mathbb{N}$, let $d_2(n)$ denote the sum of the digits in the binary expansion of n.

Proposition 1.2. *For all $n \in \mathbb{N}$, we have $t_n = d_2(n) \bmod 2$.*

Proof. Note that d_2 satisfies the following recurrence relations: $d_2(0) = 0$; $d_2(2n) = d_2(n)$; and $d_2(2n+1) = d_2(n) + 1$. Since $d_2(n) \bmod 2$ satisfies the same recurrences defining t_n, we have $t_n = d_2(n) \bmod 2$. □

Exercise 1.1. If $\boldsymbol{t} = t_0t_1t_2\cdots$ is the Thue – Morse word, show that

$$\sum_{n\geq 0}(-1)^{t_n}x^n = (1-x)(1-x^2)(1-x^4)(1-x^8)\cdots.$$

Exercise 1.2 ([AS1999])**.** Let $\boldsymbol{t} = t_0t_1t_2\cdots$ be the Thue – Morse word and let $s_n = (-1)^{t_n}$ for $n \geq 0$. Compute the following.

$$\left(\frac{1}{2}\right)^{s_0}\left(\frac{3}{4}\right)^{s_1}\left(\frac{5}{6}\right)^{s_2}\cdots\left(\frac{2i+1}{2i+2}\right)^{s_i}\cdots.$$

(*Hint*: Let $P = \prod_{n\geq 0}\big((2n+1)/(2n+2)\big)^{s_n}$, $Q = \prod_{n\geq 1}\big(2n/(2n+1)\big)^{s_n}$. Show that $PQ = Q/(2P)$.)

$$\begin{aligned} \boldsymbol{t} &= t_0\ t_1\ t_2\ t_3\ \cdots\ t_m\ \cdots\ t_{2m}\ t_{2m+1}\ \cdots \\ &= 0\ \ 1\ \ 1\ \ 0\ \cdots\ a\ \cdots\ a\ \ \ \bar{a}\ \ \cdots. \end{aligned}$$

Figure 1. The Thue – Morse word $\boldsymbol{t}$. Here $a \in \{0,1\}$.

2. The Thue – Morse morphism

Definition 1.3. The **Thue – Morse morphism** is the map $\mu\colon \{0,1\}^* \to \{0,1\}^*$ defined by $\mu(0) = 01$ and $\mu(1) = 10$.

The Thue – Morse morphism μ is an example of a 2-*uniform morphism*: a morphism ξ of words over an alphabet A is a ***k*-uniform morphism** if $\xi(a)$ is a word of length k for all $a \in A$. Section 1 of Chapter 2 will have more to say about k-uniform morphisms.

If $\boldsymbol{s}$ is an infinite word over the alphabet $\{0,1\}$, then let $\bar{\boldsymbol{s}}$ be the image of $\boldsymbol{s}$ under the endomorphism defined by $0 \mapsto 1$ and $1 \mapsto 0$. This morphism is often called the **exchange morphism**. Note that $\mu(\bar{\boldsymbol{s}}) = \overline{\mu(\boldsymbol{s})}$ for any finite or infinite word $\boldsymbol{s}$ over $\{0,1\}$.

Proposition 1.4. *The Thue – Morse word* $\boldsymbol{t}$ *is a fixed point of the Thue – Morse morphism* μ*, i.e.,* $\mu(\boldsymbol{t}) = \boldsymbol{t}$*. Moreover,* $\boldsymbol{t}$ *and* $\bar{\boldsymbol{t}}$ *are the only fixed points of* μ.

Proof. Suppose $\boldsymbol{s}$ is a binary word. Since μ maps each $a \in \{0,1\}$ to $a\bar{a}$, it follows that $\big(\mu(\boldsymbol{s})\big)_{2n} = s_n$ and $\big(\mu(\boldsymbol{s})\big)_{2n+1} = \bar{s}_n$ for all $n \geq 0$. So if $\mu(\boldsymbol{s}) = \boldsymbol{s}$, then $s_{2n} = s_n$ and $s_{2n+1} = \bar{s}_n$. If $s_0 = 0$, then $\boldsymbol{s} = \boldsymbol{t}$; and if $s_0 = 1$, then $\boldsymbol{s} = \bar{\boldsymbol{t}}$. Therefore, $\boldsymbol{t}$ and $\bar{\boldsymbol{t}}$ are the only fixed points of μ. □

The above result characterizes the Thue – Morse word as the infinite binary word beginning with 0 that is a fixed point of μ. Defining infinite words in this fashion, as fixed points of morphisms, is a useful technique that will be employed often in what follows. Let us underscore the necessary ingredients. Fix $n \in \mathbb{N}$ and a monoid endomorphism $\varphi\colon A^* \to A^*$. We write φ^n for the n-fold composition of φ with itself, the nth **iterate** of φ. If a is a prefix of $\varphi(a)$ for a given $a \in A$, then $\varphi^n(a)$ is a prefix of $\varphi^{n+1}(a)$ for all positive integers n: indeed, writing $\varphi(a) = au$, we have

$$\varphi^{n+1}(a) = \varphi^n\big(\varphi(a)\big) = \varphi^n(au) = \varphi^n(a)\varphi^n(u).$$

Therefore, the sequence $\varphi^1(a), \varphi^2(a), \varphi^3(a), \ldots$ has a (unique) well-defined limit, which we denote by

$$\varphi^\infty(a) = \lim_{n\to\infty} \varphi^n(a).$$

Not surprisingly, the Thue – Morse word is a limit of the morphism μ.

Proposition 1.5. *The Thue – Morse word* $\boldsymbol{t}$ *is the limit* $\mu^\infty(0) = \lim_{n\to\infty} \mu^n(0)$ *of the Thue – Morse morphism* μ*. Moreover,* $\bar{\boldsymbol{t}} = \mu^\infty(1)$.

Proof. Note that $\mu\big(\mu^\infty(0)\big) = \mu^\infty(0)$. Therefore, $\mu^\infty(0)$ is a fixed point of μ beginning with 0. So $\boldsymbol{t} = \mu^\infty(0)$ by Proposition 1.4. □

By formalizing the properties of the iterates μ^n of μ, we arrive at another recursive construction of the Thue – Morse word that is independent of the Thue – Morse morphism. This characterization has many interesting consequences, several of which are explored in the exercises.

Proposition 1.6. *Fix* $u_0 = 0$ *and* $v_0 = 1$*, and let* $u_{n+1} = u_n v_n$ *and* $v_{n+1} = v_n u_n$ *for* $n \geq 0$*. Then for all* $n \geq 0$*, one has:*

(i) $u_n = \mu^n(0)$ *and* $v_n = \mu^n(1)$;
(ii) $v_n = \bar{u}_n$ *and* $u_n = \bar{v}_n$;
(iii) *for* n *even,* u_n *and* v_n *are palindromes*;
(iv) *for* n *odd,* $\widetilde{u_n} = v_n$.

PROOF. Exercise 1.5. □

Exercise 1.3. If $\boldsymbol{t}$ is the Thue – Morse word and μ the Thue – Morse morphism, then $\mu(t_n) = t_{2n}t_{2n+1}$ for all $n \geq 0$.

Exercise 1.4. For any finite binary word w, $\mu(\overline{w}) = \overline{\mu(w)}$ and $\mu(\widetilde{w}) = \widetilde{\mu(w)}$.

Exercise 1.5. Prove Proposition 1.6. (*Hint*: Proceed by induction on n.)

Exercise 1.6. Show that the Thue – Morse word is a word over the alphabet $\{0110, 1001\}$.

Exercise 1.7. The set of finite factors of $\boldsymbol{t}$ is equal to the set of finite factors of $\bar{\boldsymbol{t}}$.

Exercise 1.8. If s is a finite factor of $\boldsymbol{t}$, then $\widetilde{s}$ is also a factor of $\boldsymbol{t}$.

Exercise 1.9 (Characterization of the blocks of $\boldsymbol{t}$, [AAB+1995]). Let $A = (a_n)_{n\geq 0} = 0, 1, 3, 4, 5, 7, 9, 11, 12, 13, \ldots$ denote the lexicographically least subsequence of nonnegative integers satisfying, for all $m \geq 1$, if $m \in A$, then $2m \notin A$. Show that

$$\boldsymbol{t} = 0^{a_1 - a_0} 1^{a_2 - a_1} 0^{a_3 - a_2} 1^{a_4 - a_3} \cdots .$$

Exercise 1.10 ([Pro1851]). Define an infinite word $\boldsymbol{a}$ by letting a_i (for $i \geq 1$) denote the biggest integer j such that 2^{j-1} divides i without remainder. The first few letters of $\boldsymbol{a}$ are $1213121412131215\cdots$. Define another infinite word $\boldsymbol{b}$ to be the word obtained from $(01)^\infty = 010101\cdots$ by deleting a_i letters after skipping two letters. That is, keep 2 letters, delete $a_1 = 1$ letter, keep 2 letters, delete $a_2 = 2$ letters, keep 2 letters, delete $a_3 = 1$ letter, and so on. So $\boldsymbol{b}$ begins as $01101001\cdots$. Show that $\boldsymbol{b}$ is the Thue – Morse word.

Exercise 1.11 ([AS2003, Theorem 1.7.7]). Let $\boldsymbol{t}$ be the Thue – Morse word and μ the Thue – Morse morphism. Show that if $\phi\colon \{0,1\}^* \to \{0,1\}^*$ is a morphism such that $\phi(\boldsymbol{t}) = \boldsymbol{t}$, then $\phi = \mu^n$ for some $n \geq 0$.

3. The Tarry – Escott problem

We next highlight a connection between the Thue – Morse word and a classical problem in number theory named after Gaston Tarry and Edward B. Escott in recognition of their contributions to the problem around 1910. Early results and references appear in [DB1937] and [Wri1959].

Definition 1.7 (The Tarry – Escott problem). For $m \in \mathbb{N}$ find a positive integer r and two sequences $(a_1, \ldots, a_r)$ and $(b_1, \ldots, b_r)$ of integers such that

$$\begin{aligned} a_1 + a_2 + \cdots + a_r &= b_1 + b_2 + \cdots + b_r, \\ a_1^2 + a_2^2 + \cdots + a_r^2 &= b_1^2 + b_2^2 + \cdots + b_r^2, \\ &\vdots \\ a_1^m + a_2^m + \cdots + a_r^m &= b_1^m + b_2^m + \cdots + b_r^m. \end{aligned}$$

If $(a_1, \ldots, a_r)$ and $(b_1, \ldots, b_r)$ form a solution to the Tarry – Escott problem for $m \in \mathbb{N}$, then we say r is the **size** of the solution, m is the **degree** of the solution and we write $(a_1, \ldots, a_r) \stackrel{m}{=} (b_1, \ldots, b_r)$.

Example. The sequences $(0,3,5,6)$ and $(1,2,4,7)$ satisfy

$$0^1+3^1+5^1+6^1 = 1^1+2^1+4^1+7^1 = 14,$$
$$0^2+3^2+5^2+6^2 = 1^2+2^2+4^2+7^2 = 70.$$

Therefore, $(0,3,5,6) \stackrel{2}{=} (1,2,4,7)$. This solution has size 4 and degree 2.

Eugène Prouhet was the first to provide a general-form solution to the Tarry–Escott problem (in fact, he did it 60 years prior to the work of Tarry and Escott). He solved the Tarry–Escott problem of size 2^m and degree m for every $m > 1$ by partitioning the set of integers from 0 through $2^{m+1}-1$ into two sets using the Thue–Morse word.

Theorem 1.8 (Prouhet [Pro1851]). *For every $m > 0$, there exists a solution of size 2^m to the Tarry–Escott problem of degree m.*

PROOF. Let $\boldsymbol{t}$ be the Thue–Morse word and suppose $m > 1$. For $1 \leq i \leq 2^{m+1}$, let a_i denote the index of the ith 0 in the Thue–Morse word $\boldsymbol{t}$ and let b_i denote the index of the ith 1 in $\boldsymbol{t}$. Then the sequences $(a_1,\ldots,a_{2^m})$ and $(b_1,\ldots,b_{2^m})$ form a solution to the Tarry–Escott problem of degree m. The verification of this last statement constitutes Exercise 1.12. (Alternatively, see [Wri1959].) □

Example. From the table below, we see that the indices for the first eight 0s and 1s of $\boldsymbol{t}$ are $(0,3,5,6,9,10,12,15)$ and $(1,2,4,7,8,11,13,14)$, respectively.

t_n:	0	**1**	**1**	0	**1**	0	0	**1**	**1**	0	0	**1**	0	**1**	**1**	0
n:	0	**1**	**2**	3	**4**	5	6	**7**	**8**	9	10	**11**	12	**13**	**14**	15

We leave it to the reader to verify that

$$(0,3,5,6,9,10,12,15) \stackrel{3}{=} (1,2,4,7,8,11,13,14).$$

Prouhet was in fact interested in the more general problem of partitioning the set $\{0,1,\ldots,n^{m+1}-1\}$ into n sets such that each pair of sets form a solution to the Tarry–Escott problem of degree m. We briefly describe the partitioning; the construction is illustrated in Figure 2.

Fix positive integers n and m and consider a circle with n marked spots. Write 0 next to the first spot, then write 1 next to the second spot, and so on, except that you skip one spot for every multiple of n, two spots for every multiple of n^2, etc. until $n^{m+1}-1$ is reached. We get a partition of $\{0,1,\ldots,n^{m+1}-1\}$ into n sets by considering where each integer lies on the circle.

Example. Take $n=3$ and $m=2$. Figure 2 illustrates Prouhet's decomposition of $\{0,1,\ldots,26\}$ into the three sets

$$a = (0,5,7,11,13,15,19,21,26),$$
$$b = (1,3,8,9,14,16,20,22,24),$$
$$c = (2,4,6,10,12,17,18,23,25).$$

We leave it to the reader to verify that $a \stackrel{2}{=} b$, $a \stackrel{2}{=} c$ and $b \stackrel{2}{=} c$.

Remark. Prouhet's construction defines a generalization of the Thue–Morse word to an n-letter alphabet: use the construction to partition $\mathbb{N}$ into n sets, $P_1,\ldots,P_n$; associate to each P_j a unique letter a_j; and define a word $\boldsymbol{w}$ by $w_i = a_j$ if $i \in P_j$. The word $\boldsymbol{w}$ is called the **generalized Thue–Morse word** over the

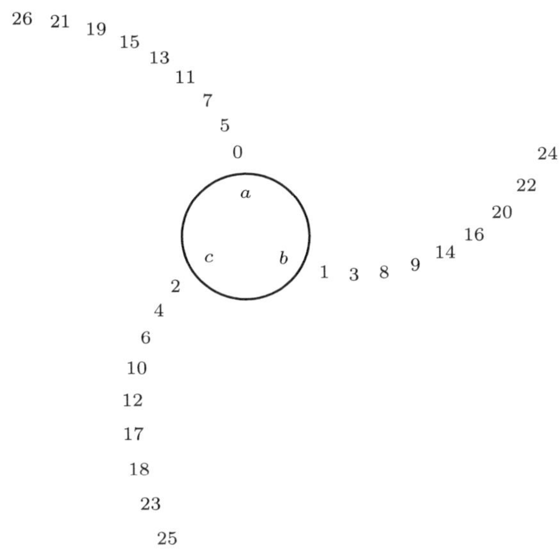

FIGURE 2. Prouhet's partitioning of $\{0, 1, \ldots, 3^3 - 1\}$.

alphabet $\{a_0, a_1, \ldots, a_{n-1}\}$. For example, the generalized Thue–Morse word over $\{0, 1, 2\}$ begins as $012120201120201\cdots$ (see Figure 2).

As with the Thue–Morse word $\boldsymbol{t}$, there are several equivalent characterizations of the generalized Thue–Morse word $\boldsymbol{w}$. It can also be constructed using n-uniform morphisms: if $\gamma(a_i) = a_i a_{i+1} \cdots a_n a_0 \cdots a_{i-1}$ for all $1 \leq i \leq n$, then $\boldsymbol{w} = \gamma^\infty(a_0)$. And if $a_i = i$ for all $0 \leq i \leq n-1$, then w_m is the sum of the digits in the base n expansion of m modulo n.

Recent research surrounding the Tarry–Escott problem includes studying the structure of solutions of minimal size and also multi-dimensional generalizations. We describe each briefly. A solution to the Tarry–Escott problem of degree m is said to be **ideal** if its size is $m + 1$. Ideal solutions are known to exist for sizes $1, 2, \ldots, 10$ and 12; see [BI1994, BLP2003].

Example. The first ideal solution of degree greater than 10 was found by Nuutti Kuosa, Jean-Charles Meyrignac and Chen Shuwen [BLP2003]:

$$(0, 11, 24, 65, 90, 129, 173, 212, 237, 278, 291, 302)$$
$$\stackrel{11}{=} (3, 5, 30, 57, 104, 116, 186, 198, 245, 272, 297, 299).$$

A multi-dimensional generalization of the Tarry–Escott problem was recently introduced by Andreas Alpers and Rob Tijdeman [AT2007]. Their results include the following generalization of Prouhet's result.

Theorem 1.9. *For every $k \in \mathbb{N}$, there exist different multisets*

$$\{(a_1, b_1), \ldots, (a_{2^k}, b_{2^k})\} \subseteq \mathbb{Z}^2 \quad \textit{and} \quad \{(c_1, d_1), \ldots, (c_{2^k}, d_{2^k})\} \subseteq \mathbb{Z}^2,$$

with $a_i \neq b_i$ for at least one $i \in \{1, 2, \ldots, 2^k\}$, such that

$$\sum_{i=1}^{2^k} a_i^{\varepsilon_1} b_i^{\varepsilon_2} = \sum_{i=1}^{2^k} c_i^{\varepsilon_1} d_i^{\varepsilon_2}$$

for all nonnegative integers ε_1 and ε_2 with $\varepsilon_1 + \varepsilon_2 \leq k$.

Exercise 1.12. Prove Theorem 1.8; that is, show that Prouhet's solution satisfies the Tarry–Escott problem.

Exercise 1.13. If $(a_1, a_2, \ldots, a_r)$ and $(b_1, b_2, \ldots, b_r)$ form an ideal solution to the Tarry–Escott problem, then the polynomials $(x - a_1)(x - a_2) \cdots (x - a_r)$ and $(x - b_1)(x - b_2) \cdots (x - b_r)$ differ only in their constant terms.

Exercise 1.14. Suppose $\{a_1, \ldots, a_r\}$ and $\{b_1, \ldots, b_r\}$ are distinct sets of integers. The following are equivalent.

(a) $\sum_{i=1}^{r} a_i^j = \sum_{i=1}^{r} b_i^j$ for $j = 1, \ldots, k$.
(b) $(x-1)^{k+1}$ divides the polynomial $\sum_{i=1}^{r}(x^{a_i} - x^{b_i})$.
(c) The degree of the polynomial $(x - a_1) \cdots (x - a_r) - (x - b_1) \cdots (x - b_r)$ is at most $r - (k+1)$.

Exercise 1.15. Suppose $(a_1, \ldots, a_r)$ and $(b_1, \ldots, b_r)$ form a solution to the Tarry–Escott problem of degree m. If λ and ν are positive integers, then

$$(\lambda a_1 + \nu, \ldots, \lambda a_r + \nu) \quad \text{and} \quad (\lambda b_1 + \nu, \ldots, \lambda b_r + \nu)$$

also form a solution of size r and degree m. (*Hint*: Use the previous exercise.)

4. Magic squares

A **magic square** of order $m \in \mathbb{N}$ is an $m \times m$ matrix whose entries are distinct elements of $\{1, 2, \ldots, m^2\}$ such that the sum of the entries in every row, column and diagonal is the same. In Exercise 1.16, it is shown that this sum must be $\frac{1}{2}m(m^2+1)$. In what follows we outline, without proof, a method to construct a magic square of order 2^m for all $m \geq 2$ using the Thue–Morse word $\boldsymbol{t}$. The reader is referred to [AL1977] for a proof.

To construct a magic square M of order 2^m for $m \geq 2$, first number the entries of M from left to right, top to bottom, beginning with 1. Let the nth entry of M be n if $t_{n-1} = 1$. Finally, arrange the unused numbers in decreasing order to fill the remaining entries of M from left to right and top to bottom.

Example. We use the above method to construct a magic square of order $2^2 = 4$. Consider the first $4^2 = 16$ letters of the Thue–Morse word.

n:	1	2	3	4	5	6	7	8	9	10	11	12	13	14	15	16
t_{n-1}:	0	**1**	**1**	0	**1**	0	0	**1**	**1**	0	0	**1**	0	**1**	**1**	0

From the above table, we see that $t_{n-1} = 1$ if $n \in \{2, 3, 5, 8, 9, 12, 14, 15\}$, so we obtain the partially-completed magic square at the left in Figure 3. The unused numbers, in decreasing order, are $16, 13, 11, 10, 7, 6, 4, 1$. These are used to fill in the empty entries, preserving the order. The resulting magic square is shown at the right in Figure 3.

Remark. The magic square in Figure 3, with the central columns interchanged, appears in the engraving *Melencolia* I (Figure 5) by the German Renaissance artist and mathematician Albrecht Dürer. A similar "magic square", depicted in Figure 4, appears on a façade of *La Sagrada Família*, a basilica in Barcelona, Spain. It is obtained from Dürer's magic square by subtracting 1 from four cells so that the sum of the rows, columns and diagonals is 33. Strictly speaking, it is not a magic square because there are two occurrences of 10 and 14 and it does not include 12 or 16.

	2	3	
5			8
9			12
	14	15	

16	2	3	**13**
5	**11**	**10**	8
9	**7**	**6**	12
4	14	15	**1**

FIGURE 3. Left: the nth box of the magic square is n if $t_{n-1} = 1$. Right: the unused numbers (in **boldface**) are inserted in decreasing order into the empty boxes.

1	14	14	4
11	7	6	9
8	10	10	5
13	2	3	15

FIGURE 4. This "magic square" appears on the Passion façade of *La Sagrada Família*, a basilica in Barcelona, Spain. The basilica was originally designed by the architect Antoni Gaudí (1852–1926).

Exercise 1.16. If M is a magic square of order m, prove that the sum of every column, row and diagonal is equal to $\frac{1}{2}m(m^2+1)$.

Exercise 1.17. Construct a magic square of order 8.

FIGURE 5. *Melencolia* I by Albrecht Dürer.

CHAPTER 2

Combinatorics of the Thue – Morse Word

This chapter uses the Thue – Morse word to shed light on several classical and modern results on the combinatorics of words.

1. Automatic sequences

We begin by presenting a characterization of the Thue – Morse word using automata. Briefly, an automaton is a model of computation; it accepts as input a finite word and uses the letters of the word to transition from state to state. Automata are used in this section to construct a class of infinite words called automatic sequences and in Section 5.1 to construct certain languages.

Definition 2.1. A **finite deterministic automaton** $\mathcal{A} = \langle A, Q, q_0, F, \cdot \rangle$ consists of an alphabet A, a finite set Q of **states**, an **initial state** q_0, a set $F \subseteq Q$ of **final states**, and a **next state function** $\cdot : Q \times A \to Q$.

For the empty word ϵ and each state $q \in Q$, we define $q \cdot \epsilon = q$. For any $u \in A^*$ and $a \in A$, we define $q \cdot ua = (q \cdot u) \cdot a$. This extends the domain of the next state function to $Q \times A^*$.

A finite deterministic automaton $\mathcal{A} = \langle A, Q, q_0, F, \cdot \rangle$ can be represented as an adorned directed graph as follows (see Figure 1 and Figure 2 for examples): the vertex set of the graph corresponds to the set of states Q, with each vertex labelled by the corresponding state; there is a labelled arrow $q \xrightarrow{a} p$ if and only if $q \cdot a = p$, where $a \in A$ and $q, p \in Q$; there is an (unlabelled) arrow pointing at the initial state q_0; and the final states are represented by a double circle. Note that if $q \in Q$ and $a_1, \ldots, a_i \in A$, then computing $q \cdot (a_1 \cdots a_i)$ amounts to starting at the vertex q and following the unique edge $q \xrightarrow{a_1} q_1$ starting at q and labelled a_1, then following the unique edge $q_1 \xrightarrow{a_2} q_2$, and so on. The last vertex of this path is the state $q \cdot (a_1 \cdots a_i)$.

The Thue – Morse word $\boldsymbol{t}$ is obtained from the automaton of Figure 1 as follows. For each $n \in \mathbb{N}$, let $\operatorname{bin}(n) \in \{0, 1\}^*$ denote the binary expansion of n and consider the state $s_n = a \cdot \operatorname{bin}(n)$. If $\operatorname{bin}(n)$ has an even number of 1s, then $s_n = a$; otherwise $s_n = b$. Since $t_n = d_2(n) \bmod 2$, where $d_2(n)$ denotes the number of 1s occurring

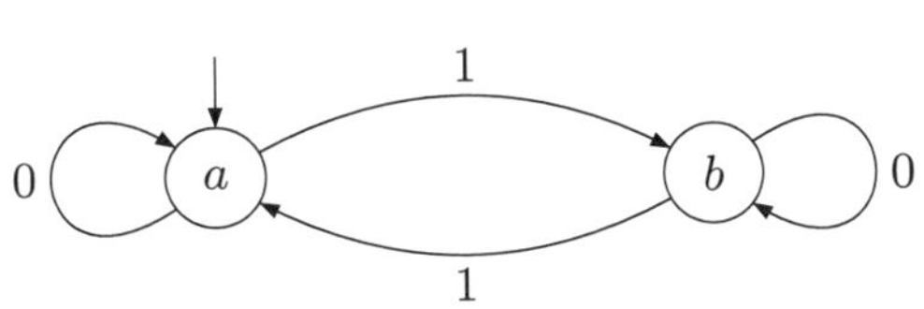

FIGURE 1. An automaton with states $Q = \{a, b\}$, alphabet $A = \{0, 1\}$ and initial state $q_0 = a$. There are no final states.

in bin(n) (Proposition 1.2), it follows that $\phi(s_n) = t_n$, where ϕ is the morphism defined by $\phi(a) = 0$ and $\phi(b) = 1$. This construction is a realization of $\boldsymbol{t}$ as an *automatic sequence.*

Definition 2.2. Let $\langle \Sigma_k, Q, q_0, F, \cdot \rangle$ be a finite deterministic automaton over the alphabet $\Sigma_k = \{0, 1, \dots, k-1\}$ for some $k \in \mathbb{N}$ and let $\phi\colon Q \to X$ denote a function from the set of states into some alphabet X. A ***k*-automatic sequence** over X is the infinite word $x_0x_1x_2\cdots$ over the alphabet X given by defining $x_n = \phi(q_0 \cdot a_0a_1\cdots a_i)$, where $a_0a_1\cdots a_i \in \Sigma_k^*$ is the base-k expansion of $n \in \mathbb{N}$.

Proposition 2.3. *The Thue–Morse word is a 2-automatic sequence.*

Recall that a morphism $\xi\colon A^* \to A^*$ is k-uniform for some $k \in \mathbb{N}$ if the word $\xi(a)$ has length k for all $a \in A$. Alan Cobham proved that k-automatic sequences correspond to 1-uniform morphic images of fixed points of k-uniform morphisms [Cob1972]. The following theorem is one half of Cobham's result; see [Cob1972] or Theorem 6.3.2 of [AS2003] for the complete statement.

Theorem 2.4 (Cobham). *If an infinite word $\boldsymbol{w}$ is the fixed point of a k-uniform morphism for some $k \geq 2$, then $\boldsymbol{w}$ is a k-automatic sequence.*

Proof. Suppose $\boldsymbol{w}$ is a word over A and let ξ denote a k-uniform morphism with $\xi(\boldsymbol{w}) = \boldsymbol{w}$. Define an automaton over the alphabet $\{0, 1, \dots, k-1\}$ with state set A, initial state w_0 and next state function given by defining, for $0 \leq i < k$, $q \cdot i$ to be the letter in position i of $\xi(q)$. Then induction on n establishes that $w_0 \cdot (a_0 \cdots a_i) = w_n$, where $a_0 \cdots a_i$ is the base-k expansion of n (Exercise 2.1). $\square$

By Proposition 1.4, the Thue–Morse word is the fixed point of the 2-uniform Thue–Morse morphism μ. Accordingly, this result gives a second proof that the Thue–Morse word is 2-automatic (in fact, it rebuilds the automaton in Figure 1).

There are several interesting connections between automatic sequences and other areas of mathematics and the physical sciences; for example, several important sequences occurring in number theory are k-automatic sequences. These connections are described in the book by Jean-Paul Allouche and Jeffrey Shallit [AS2003].

Remark. Evidently, the automata formalism allows one to succinctly describe words with complicated factor-behaviour. But it has many "practical" applications as well. The reader is invited to reflect on the following question before returning to our discussion on the Thue–Morse word.

> Is it possible, given two positive integers k and n, to build a word w over an n-element alphabet A such that every word of length k over A occurs in w exactly once?

For example, 10011 possesses 00, 01, 10 and 11 as factors, each occurring exactly once. Such words, when they exist, are called **linear de Bruijn words**. See Exercise 2.3 for more details on their construction. Also, see [Lem1970] and [Men2003] for applications to shift registers and fractal rendering, respectively, and [Mor2004] for connections to Part I of this book (specifically, to Lyndon words).

Exercise 2.1. Complete the proof of Theorem 2.4.

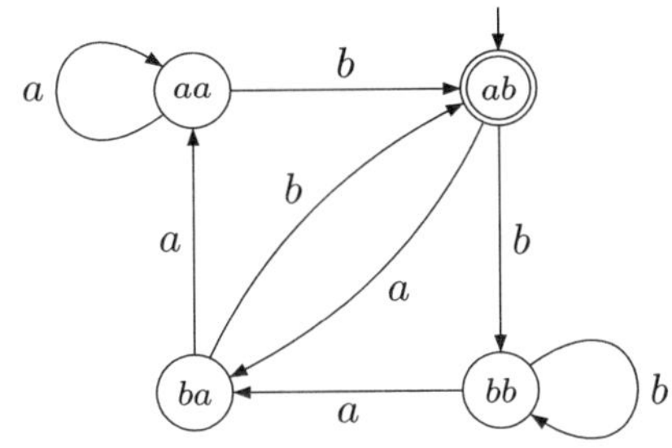

FIGURE 2. An automaton building a de Bruijn word $w_{3,2}$.

Exercise 2.2. A **(circular) de Bruijn word** $w_{(k,n)}$ is a word over an n element alphabet A such that every length k word over A appears as a factor of the circular word $(w_{(k,n)})$ exactly once.

(a) Use the automaton pictured in Figure 2 to prove that there exists a de Bruijn word $w_{(3,2)}$.

(b) Prove that de Bruijn words $w_{(k,2)}$ exist for all positive integers k. (*Hint*: The states in Figure 2 are members of $\{a,b\}^{3-1}$ and the edges are of the form $yu \xrightarrow{x} ux$ for $x, y \in \{a,b\}$.)

(c) Prove that de Bruijn words exist for all positive integers k and n. (*Hint*: The algorithm you will give should construct a word of length n^k. The minimum possible length is achieved.)

Exercise 2.3. A **linear de Bruijn word** $w_{[k,n]}$ is a word over an n element alphabet A such that every length k word over A is a factor of $w_{[k,n]}$ exactly once. Use Exercise 2.2 to prove that linear de Bruijn words $w_{[k,n]}$ exist for all positive integers k and n. (*Hint*: The minimum-length linear de Bruijn words have length $n^k + (k-1)$.)

2. Generating series

We next present a characterization of the Thue–Morse word $\boldsymbol{t}$ in terms of a generating series over $\mathbb{Z}/2\mathbb{Z}$ for the elements of $\boldsymbol{t}$. We begin by recalling the relevant notions; see also Chapters 4 and 6 of [Sta1999] or [AS2003].

A **generating series** of a sequence $(s_n)_{n\in\mathbb{N}}$ of elements of a field $\Bbbk$ is the formal power series $\sum_{n\in\mathbb{N}} s_n x^n$ in one variable x. The ring of all formal power series in the variable x and with coefficients in $\Bbbk$ is denoted by $\Bbbk[\![x]\!]$. A series $f(x) \in \Bbbk[\![x]\!]$ is said to be **rational** if there exist polynomials $p(x), q(x) \in \Bbbk[x]$ such that $f(x) = p(x)/q(x)$. An element $f(x) \in \Bbbk[\![x]\!]$ is said to be **algebraic** over the quotient field $\Bbbk(x)$ of the polynomial ring $\Bbbk[x]$ if there exist $p_0(x), p_1(x), \ldots, p_n(x) \in \Bbbk(x)$, not all 0, such that

$$p_0(x) + p_1(x)f(x) + \cdots + p_n(x)\big(f(x)\big)^n = 0.$$

If $f(x) \in \Bbbk[\![x]\!]$ is not algebraic over $\Bbbk(x)$, then $f(x)$ is **transcendental** over $\Bbbk(x)$. Note that if $f(x) \in \Bbbk[\![x]\!]$ is rational, then it is algebraic over $\Bbbk(x)$.

Example. The prototypical example of a rational generating series is the series $F(x) = \sum_{n\geq 0} F_n x^n$ built from the **Fibonacci numbers** ($F_0 = 1, F_1 = 1$; $F_n = F_{n-1} + F_{n-2}$ for $n \geq 2$). One has

$$1 + (x^2 + x - 1)F(x) = 0.$$

Let $\boldsymbol{t}(x)$ denote the generating series for the letters of the Thue–Morse word $\boldsymbol{t}$ over the finite field $\mathbb{Z}/2\mathbb{Z}$ of two elements. That is,

$$\boldsymbol{t}(x) = \sum_{n\geq 0} t_n x^n \in (\mathbb{Z}/2\mathbb{Z})[\![x]\!].$$

Also let $\bar{\boldsymbol{t}}(x) \in (\mathbb{Z}/2\mathbb{Z})[\![x]\!]$ denote the generating series for the letters of $\bar{\boldsymbol{t}}$. The next result shows that $\boldsymbol{t}(x)$ and $\bar{\boldsymbol{t}}(x)$ are algebraic over $(\mathbb{Z}/2\mathbb{Z})(x)$.

Proposition 2.5 ([CKMR1980]). *The generating series $\boldsymbol{t}(x)$ and $\bar{\boldsymbol{t}}(x)$ are the two solutions to the equation*

$$x + (1+x)^2 Z + (1+x)^3 Z^2 = 0.$$

PROOF. We prove only that $\boldsymbol{t}(x)$ is a solution. Observe that $f(x)^2 = f(x^2)$ for any formal power series $f \in (\mathbb{Z}/2\mathbb{Z})[\![x]\!]$. Thus,

$$\begin{aligned}
\boldsymbol{t}(x) &= \sum_{n\geq 0} t_{2n} x^{2n} + \sum_{n\geq 0} t_{2n+1} x^{2n+1} \\
&= \sum_{n\geq 0} t_n x^{2n} + \sum_{n\geq 0} (1+t_n) x^{2n+1} \\
&= \sum_{n\geq 0} t_n x^{2n} + \sum_{n\geq 0} x^{2n+1} + \sum_{n\geq 0} t_n x^{2n+1} \\
&= \boldsymbol{t}(x^2) + \frac{x}{1+x^2} + x\boldsymbol{t}(x^2) \\
&= (1+x)\boldsymbol{t}(x)^2 + \frac{x}{1+x^2}.
\end{aligned}$$

Therefore, $(1+x)^2\boldsymbol{t}(x) = (1+x)^3\boldsymbol{t}(x)^2 + x$. □

Let $\mathbb{F}_q$ denote the finite field with q elements, q a power of a prime. A theorem of Harry Furstenberg [Fur1967] states that over $\mathbb{F}_q$, every algebraic series in one variable is the *diagonal* of a rational series in two variables, where the **diagonal** of a series $\sum_{n,m\in\mathbb{N}} a(n,m)x^n y^m$ is defined to be $\sum_{n\in\mathbb{N}} a(n,n)x^n$ (see also [AS2003, Theorem 12.7.3]). Jean-Paul Allouche noticed that the Thue–Morse series $\boldsymbol{t}(x)$ is the diagonal of the following rational series in $\mathbb{F}_2(x)$ (see Exercise 2.4):

$$R(x,y) = \frac{y}{1 + y(1+xy) + \dfrac{x}{(1+xy)^2}}.$$

Open Question. *Find a combinatorial reason for this identity.*

In [CKMR1980], Gilles Christol, Teturo Kamae, Michel Mendès France, and Gérard Rauzy classify automatic sequences consisting of elements from a finite field in terms of the algebraicity of their generating series. Explicitly, they show that a sequence $(s_n)_{n\in\mathbb{N}}$ over $\mathbb{F}_q$ is q-automatic if and only if $\sum_{n\in\mathbb{N}} s_n x^n$ is algebraic over $\mathbb{F}_q(x)$. Together with the above computation, this gives another proof that the Thue–Morse word is a 2-automatic sequence.

Exercise 2.4. Show that the diagonal $D(x) \in (\mathbb{Z}/2\mathbb{Z})[\![x]\!]$ of the series

$$R(x,y) = \cfrac{y}{1 + y(1+xy) + \cfrac{x}{(1+xy)^2}}.$$

satisfies $(1+x)^3 D(x)^2 + (1+x)^2 D(x) + x = 0$. Conclude that $D(x) = \boldsymbol{t}(x)$.

3. Overlaps

An **overlap** is a word of the form $auaua$, where a is a letter and u is a (possibly empty) word. A word w is said to be **overlap-free** if no overlap is a factor of w. The following result was first proved by Axel Thue [Thu1912].

Theorem 2.6. *The Thue – Morse word* $\boldsymbol{t}$ *is overlap-free.*

We need the following two lemmas.

Lemma 2.7. *Let* $C = \{01, 10\}$. *If* $x \in C^*$, *then* $0x0 \notin C^*$ *and* $1x1 \notin C^*$.

PROOF (Robert Cori). If $w \in C^*$, then $|w|_0 = |w|_1$. Since $x \in C^*$, we have $|0x0|_0 = |x|_0 + 2 = |x|_1 + 2 > |x|_1 = |0x0|_1$. Thus $0x0 \notin C^*$. □

Lemma 2.8. *Fix* $w \in \{0,1\}^*$ *and let* μ *denote the Thue – Morse morphism. If* w *is overlap-free, then* $\mu(w)$ *is overlap-free.*

PROOF. Let w be a shortest word such that $\mu(w)$ is not overlap-free. So there exist words $x, y, u \in \{0,1\}^*$ and a letter $a \in \{0,1\}$ such that $\mu(w) = xauauay$. Since μ is 2-uniform, the minimality of w implies that $|x|, |y| \leq 1$. We consider two cases.

Case $|x| = 1$. Here $|y| = 0$, since $|auaua|$ is odd. Also, $x \neq a$ since aa is not in the image of μ. Hence, $\mu(w) = \bar{a}auaua$. If $|u|$ is even then both u and aua are in $\{01, 10\}^*$, contradicting the previous lemma. So $|u|$ is odd. Hence, ua is in the image of μ; that is, there exists some $v \in \{0,1\}^*$ such that $ua = \mu(v\bar{a})$. Since $\mu(w) = \bar{a}a\mu(v\bar{a})\mu(v\bar{a}) = \mu(\bar{a}v\bar{a}v\bar{a})$ and μ is injective, we have $w = \bar{a}v\bar{a}v\bar{a}$. So w is not overlap-free.

Case $|x| = 0$. Here $y = \bar{a}$, and the preceding argument again shows that w is not overlap-free. □

PROOF OF THEOREM 2.6. By Lemma 2.8, $\mu^n(0)$ is overlap-free, so the prefixes of $\mu^\infty(0)$ are overlap-free. Hence, $\boldsymbol{t}$ is overlap-free by Proposition 1.5. □

Remark. Since the Thue – Morse word $\boldsymbol{t}$ is overlap-free and a fixed point of a nonidentity morphism (Proposition 1.4), it is natural ask which infinite binary words have these properties. This was answered by Patrice Séébold in [Séé1982]. (See also [BS1993] and [AS2003, Corollary 1.7.9].) Remarkably, $\boldsymbol{t}$ and $\bar{\boldsymbol{t}}$ are the only infinite binary words with these properties.

Exercise 2.5. Show that arbitrarily long squares can be found as factors of the Thue – Morse word $\boldsymbol{t}$.

Exercise 2.6 ([Brl1989]). Let $\boldsymbol{t}$ be the Thue – Morse word, and consider a factorization $\boldsymbol{t} = x\boldsymbol{s}$, where x is a nonempty finite word and $\boldsymbol{s}$ is an infinite word. Then either x ends with a square or $\boldsymbol{s}$ starts with a square.

Exercise 2.7 ([Ber1995]). The Thue – Morse word is the lexicographically greatest infinite overlap-free binary word beginning with 0.

4. Complexity

The **complexity function** $c_w(n)$ of a word w is the function that counts the number of distinct factors of length n in the word w. Closed-form expressions for the complexity function of the Thue–Morse word were first discovered in 1989 independently by Srečko Brlek [Brl1989] and Aldo de Luca and Stefano Varricchio [dLV1989]. In 1995, John Tromp and Jeffrey Shallit provided a new approach [TS1995] that recovers the earlier results. Our presentation is based on the work of de Luca and Varricchio [dLV1989].

Lemma 2.9. *Let u be a factor of the Thue–Morse word $\boldsymbol{t}$ of length at least four. Then the starting position of two occurrences of u have the same parity.*

PROOF. Suppose u begins by $t_n t_{n+1} t_{n+2} t_{n+3} = t_m t_{m+1} t_{m+2} t_{m+3}$ with n even and m odd. Since n is even, u must begin by $a\bar{a}$ for some $a \in \{0,1\}$ (e.g., by Definition 1.1). Thus $t_{m+1} = t_{n+1} = \bar{a}$. Since $m+1$ is even and $\boldsymbol{t}$ is a word over the alphabet $\{01, 10\}$ (Exercise 1.6), we have $t_{n+2} = t_{m+2} = a$. Since $n+2$ is even, $t_{m+3} = t_{n+3} = \bar{a}$. Since $m+3$ is even, $t_{m+4} = a$. Therefore, $t_m \cdots t_{m+4} = a\bar{a}a\bar{a}a$ is an overlap as well as a factor of $\boldsymbol{t}$, contradicting the fact that $\boldsymbol{t}$ is overlap-free. □

As a consequence we obtain a recursive definition of the complexity function $c_{\boldsymbol{t}}(n)$ of the Thue–Morse word $\boldsymbol{t}$.

Proposition 2.10. $c_{\boldsymbol{t}}(0) = 1$, $c_{\boldsymbol{t}}(1) = 2$, $c_{\boldsymbol{t}}(2) = 4$, $c_{\boldsymbol{t}}(3) = 6$, *and for* $m \geq 2$,

$$c_{\boldsymbol{t}}(2m+1) = 2c_{\boldsymbol{t}}(m+1) \quad \text{and} \quad c_{\boldsymbol{t}}(2m) = c_{\boldsymbol{t}}(m+1) + c_{\boldsymbol{t}}(m).$$

PROOF. We prove only $c_{\boldsymbol{t}}(2m+1) = 2c_{\boldsymbol{t}}(m+1)$, the argument for the other identity being similar. Let u be a factor of $\boldsymbol{t}$ of length $2m+1$ with $m \geq 1$. We consider two cases.

If u begins at an even position, then the letter following u in $\boldsymbol{t}$ is determined by the last letter of u since $\boldsymbol{t}$ is a word over the alphabet $\{01, 10\}$. Therefore, there is a bijection between the factors of length $2m+1$ that begin at an even position and the factors of length $2m+2$ that begin at an even position. The latter are in bijection with factors of length $m+1$ since $t_{2n} = t_n$ and $t_{2n+1} = \bar{t}_n$ for all $n \geq 0$. Therefore, there are $c_{\boldsymbol{t}}(m+1)$ factors of $\boldsymbol{t}$ of length $2m+1$ that begin at an even position.

Similarly, if u begins at an odd position then the letter preceding u in $\boldsymbol{t}$ is determined by the first letter of u, so there is a bijection between factors of length $2m+1$ beginning in an odd position and factors of length $m+1$. Therefore, there are $c_{\boldsymbol{t}}(m+1)$ factors of $\boldsymbol{t}$ of length $2m+1$ that begin at an odd position.

By Lemma 2.9, no factor of $\boldsymbol{t}$ of length at least 4 can begin at both an odd position and an even position, so $c_{\boldsymbol{t}}(2m+1) = 2c_{\boldsymbol{t}}(m+1)$. □

Our next aim is a closed-form expression for the complexity function $c_{\boldsymbol{t}}(n)$ of the Thue–Morse word. A (finite) factor u of the Thue–Morse word $\boldsymbol{t}$ is said to be **right special** if both $u0$ and $u1$ are also factors of $\boldsymbol{t}$. (The **left special** factors of $\boldsymbol{t}$

n:	0	1	2	3	4	5	6	7	8	9	10	11
$c_{\boldsymbol{t}}(n)$:	1	2	4	6	10	12	16	20	22	24	28	32

FIGURE 3. The first 12 values of the complexity function $c_{\boldsymbol{t}}(n)$.

n:	0	1	2	3	4	5	6	7	8	9	10	11
$c_{\boldsymbol{t}}(n)$:	1	2	4	6	10	12	16	20	22	24	28	32
$s_{\boldsymbol{t}}(n)$:	1	2	2	4	2	4	4	2	2	4	4	4

FIGURE 4. The first 12 values of $c_{\boldsymbol{t}}(n)$ and $s_{\boldsymbol{t}}(n)$.

are defined analogously.) Let $s_{\boldsymbol{t}}(n)$ denote the number of right special factors of $\boldsymbol{t}$ of length n. There is a strong connection between the factors of $\boldsymbol{t}$ and its right special factors, which we exploit to develop a closed-form expression for $c_{\boldsymbol{t}}(n)$. Each right special factor of length n determines two distinct factors of length $n+1$ while any factor that is not right special determines only one factor of length $n+1$. Thus, $c_{\boldsymbol{t}}$ and $s_{\boldsymbol{t}}$ are related by

$$c_{\boldsymbol{t}}(n+1) = 2s_{\boldsymbol{t}}(n) + \big(c_{\boldsymbol{t}}(n) - s_{\boldsymbol{t}}(n)\big) = s_{\boldsymbol{t}}(n) + c_{\boldsymbol{t}}(n). \tag{2.11}$$

Proposition 2.12. $s_{\boldsymbol{t}}(1) = 2$, $s_{\boldsymbol{t}}(2) = 2$, $s_{\boldsymbol{t}}(3) = 4$ *and for all* $m \geq 2$,

$$s_{\boldsymbol{t}}(2m+1) = s_{\boldsymbol{t}}(m+1) \quad \textit{and} \quad s_{\boldsymbol{t}}(2m) = s_{\boldsymbol{t}}(m).$$

PROOF. This follows from (2.11) and the recursion for $c_{\boldsymbol{t}}(n)$ developed in Proposition 2.10: if $m \geq 2$, then

$$\begin{aligned} s_{\boldsymbol{t}}(2m+1) &= c_{\boldsymbol{t}}(2m+2) - c_{\boldsymbol{t}}(2m+1) \\ &= \big(c_{\boldsymbol{t}}(m+2) + c_{\boldsymbol{t}}(m+1)\big) - 2c_{\boldsymbol{t}}(m+1) \\ &= c_{\boldsymbol{t}}(m+2) - c_{\boldsymbol{t}}(m+1) \\ &= s_{\boldsymbol{t}}(m+1), \end{aligned}$$

and

$$\begin{aligned} s_{\boldsymbol{t}}(2m) &= c_{\boldsymbol{t}}(2m+1) - c_{\boldsymbol{t}}(2m) \\ &= 2c_{\boldsymbol{t}}(m+1) - \big(c_{\boldsymbol{t}}(m+1) + c_{\boldsymbol{t}}(m)\big) \\ &= c_{\boldsymbol{t}}(m+1) - c_{\boldsymbol{t}}(m) \\ &= s_{\boldsymbol{t}}(m). \end{aligned}$$

□

It follows immediately that $s_{\boldsymbol{t}}(n) \in \{2, 4\}$ for all $n > 0$. But we can be much more precise: for all $n \geq 3$,

$$s_{\boldsymbol{t}}(n) = \begin{cases} 4, & \text{if } n \in \bigcup_{k \in \mathbb{N}, k \geq 1} (2^k, 2^k + 2^{k-1}], \\ 2, & \text{if } n \in \bigcup_{k \in \mathbb{N}, k \geq 1} (2^k + 2^{k-1}, 2^{k+1}]. \end{cases} \tag{2.13}$$

This follows from the above recurrences for $s_{\boldsymbol{t}}(n)$. See Exercise 2.11.

Proposition 2.14. *For all* $n \geq 3$,

$$c_{\boldsymbol{t}}(n) = \begin{cases} 3 \cdot 2^k + 4(r-1), & \textit{if } 1 \leq r \leq 2^{k-1}, \\ 4 \cdot 2^k + 2(r-1), & \textit{if } 2^{k-1} < r \leq 2^k, \end{cases}$$

where k and r are uniquely determined by $n = 2^k + r$ with $1 \leq r \leq 2^k$.

PROOF. Fix $n \geq 3$. Suppose first that $n \in (2^k, 2^k + 2^{k-1}]$ for some positive integer $k \geq 1$. Since $c_{\boldsymbol{t}}(n) = 2 + \sum_{i=1}^{n-1} s_{\boldsymbol{t}}(i)$ and $s_{\boldsymbol{t}}(i) \in \{2, 4\}$ for all $i \geq 1$ (see (2.11) and (2.13), respectively), it follows that $c_{\boldsymbol{t}}(n) = 2 + 4(n-1) - 2m$, where m is the number of elements i in $\{1, 2, \ldots, n-1\}$ such that $s_{\boldsymbol{t}}(i) = 2$.

By (2.13), m is the cardinality of the set $\{1,2\} \cup \bigcup_{j=1}^{k-1}(2^j + 2^{j-1}, 2^{j+1}]$. Thus $m = 2 + (1 + 2 + \cdots + 2^{k-2}) = 2^{k-1} + 1$, and so $c_{\boldsymbol{t}}(n) = 4(n-1) - 2^k$.

If $n \in (2^k + 2^{k-1}, 2^{k+1}]$ for some positive integer $k \geq 1$, then a similar argument shows that $c_{\boldsymbol{t}}(n) = 2(n-1) + 2^{k+1}$. We conclude that

$$c_{\boldsymbol{t}}(n) = \begin{cases} 4(n-1) - 2^k, & \text{if } 2^k + 1 \leq n \leq 2^k + 2^{k-1}, \\ 2(n-1) + 2^{k+1}, & \text{if } 2^k + 2^{k-1} < n \leq 2^{k+1}, \end{cases}$$

for all $n \geq 3$, where k is a positive integer such that $2^k + 1 \leq n \leq 2^{k+1}$. Replacing n by $2^k + r$, where $r = n - 2^k$, establishes the proposition. □

Remarks.

(1) A word $\boldsymbol{s}$ is said to be **recurrent** if every finite factor of $\boldsymbol{s}$ occurs infinitely often in $\boldsymbol{s}$. Exercise 2.8 establishes that the Thue–Morse word $\boldsymbol{t}$ is recurrent. This implies that $\boldsymbol{t}$ and every suffix of $\boldsymbol{t}$ have the same complexity function, and so too does any infinite word with the same set of factors as $\boldsymbol{t}$. Surprisingly, if a recurrent infinite word $\boldsymbol{s}$ has the same complexity function as $\boldsymbol{t}$, then the set of factors of $\boldsymbol{s}$ is either the set of factors of $\boldsymbol{t}$ or the set of factors of $\delta(\boldsymbol{t})$, where δ is the letter-doubling morphism defined by $\delta(0) = 00$ and $\delta(1) = 11$ [ABG2007].

(2) A word $\boldsymbol{s}$ is said to be **uniformly recurrent** if for every $n \in \mathbb{N}$, there exists a smallest integer $R_{\boldsymbol{s}}(n)$ such that any factor of $\boldsymbol{s}$ of length n is a factor of any factor of $\boldsymbol{s}$ of length $R_{\boldsymbol{s}}(n)$. The function $R_{\boldsymbol{s}} \colon \mathbb{N} \to \mathbb{N}$ is called the **recurrence index** of $\boldsymbol{s}$. The Thue–Morse word $\boldsymbol{t}$ is uniformly recurrent with $R_{\boldsymbol{t}}(1) = 3$ because $\boldsymbol{t}$ is overlap-free and $R_{\boldsymbol{t}}(2) = 7$ because 00 is not a factor of $t_0 \cdots t_5 = 011010$ (Exercise 2.10).

The notion of special factors has come to play an important role in the theory of words. We close this section with two results concerning the left special factors of the Thue–Morse word $\boldsymbol{t}$. (Recall that u is a left special factor of $\boldsymbol{t}$ if $0u$ and $1u$ are factors of $\boldsymbol{t}$.)

Proposition 2.15. *A word u starting with 0 is a left special factor of the Thue–Morse word $\boldsymbol{t}$ if and only if it is a prefix of $\mu^n(010)$ for some $n \in \mathbb{N}$.*

PROOF. Suppose u is a prefix of $\mu^n(010)$ for some $n \in \mathbb{N}$. We use Exercise 2.13: the image under μ of a left special factor of $\boldsymbol{t}$ is a left special factor of $\boldsymbol{t}$. Since 010 is a left special factor of $\boldsymbol{t}$, we infer that $\mu^n(010)$ is a left special factor of $\boldsymbol{t}$ for all $n \geq 0$. Finally, u is a left special factor of $\boldsymbol{t}$ since it is a prefix of a left special factor of $\boldsymbol{t}$.

We show that the prefixes of $\mu^n(010)$ exhaust all the left special factors of $\boldsymbol{t}$ that begin with 0. Since the set of finite factors of $\boldsymbol{t}$ is closed under reversal (Exercise 1.8), any left special factor determines a right special factor, and conversely. Thus, the number of left special factors of $\boldsymbol{t}$ having length ℓ is equal to $s_{\boldsymbol{t}}(\ell)$. And since u is a left special factor if and only if $\bar{u}$ is a left special factor (Exercise 2.12), the number of left special factors of length ℓ that begin with 0 is equal to $\frac{1}{2}s_{\boldsymbol{t}}(\ell) \in \{1,2\}$. Since the prefixes of $\mu^n(010)$ are left special factors, we need only show that if $s_{\boldsymbol{t}}(\ell) = 4$, then there are two distinct words that appear as length ℓ prefixes of the words $\mu^n(010)$.

If $s_{\boldsymbol{t}}(\ell) = 4$, then $2^k < \ell \leq 2^k + 2^{k-1}$ for some positive integer k. The length ℓ prefix of $\mu^{k-1}(010)$ is $\mu^{k-1}(01)v$ for some nonempty prefix v of $\mu^{k-1}(0)$. The length ℓ prefix of $\mu^k(010) = \mu^{k-1}(011001)$ is $\mu^{k-1}(01)u$ for some nonempty prefix

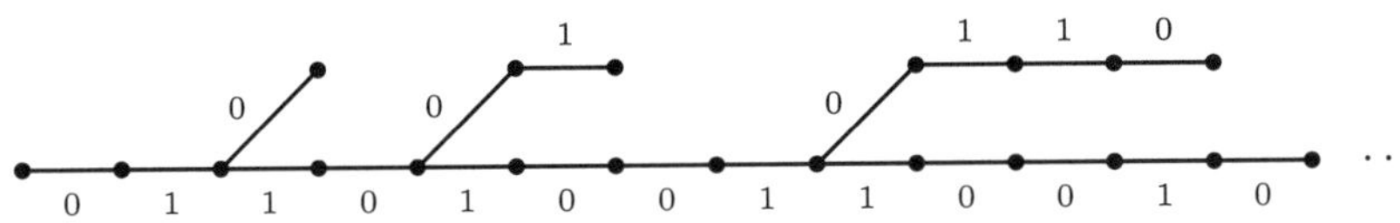

FIGURE 5. The tree of left special factors beginning with 0 of the Thue–Morse word.

u of $\mu^{k-1}(1)$. Since the first letters of u and v are different, we have at least two distinct prefixes of the words $\mu^n(010)$ $(n \geq 0)$ that have length ℓ. $\square$

Figure 5 depicts the tree of all the left special factors of the Thue–Morse word that begin with 0. It is obtained by considering the prefixes of the iterates $\mu^n(010)$ for $n \in \mathbb{N}$. Since every prefix of the Thue–Morse word is a prefix of $\mu^n(010)$ for sufficiently large n, we obtain the following immediate corollary.

Corollary 2.16. *Every prefix of the Thue–Morse word is a left special factor.*

Let $\boldsymbol{t}$ denote the Thue–Morse word and μ the Thue–Morse morphism.

Exercise 2.8. Prove that every factor of $\boldsymbol{t}$ occurs infinitely often in $\boldsymbol{t}$ (that is, prove that $\boldsymbol{t}$ is recurrent) and conclude that if $\boldsymbol{s}$ is a suffix of $\boldsymbol{t}$, then $c_{\boldsymbol{s}}(n) = c_{\boldsymbol{t}}(n)$ for all $n \in \mathbb{N}$.

Exercise 2.9 **([CH1973]).** Prove that a word $\boldsymbol{s}$ is uniformly recurrent if and only if every factor w of $\boldsymbol{s}$ occurs in $\boldsymbol{s}$ infinitely often and the distances between consecutive occurrences of w in $\boldsymbol{s}$ are bounded.

Exercise 2.10. Show that $\boldsymbol{t}$ is uniformly recurrent and the recurrence index $R_{\boldsymbol{t}}$ of $\boldsymbol{t}$ satisfies $R_{\boldsymbol{t}}(1) = 3$ and $R_{\boldsymbol{t}}(2) = 7$. (*Hint*: Use the previous exercise.)

Exercise 2.11. Suppose $n \geq 3$ and k is a positive integer such that $2^k + 1 \leq n \leq 2^{k+1}$. Then

$$s_{\boldsymbol{t}}(n) = \begin{cases} 4, & \text{if } n \in (2^k, 2^k + 2^{k-1}], \\ 2, & \text{if } n \in (2^k + 2^{k-1}, 2^{k+1}]. \end{cases}$$

(*Hint*: Proceed by induction on k using Proposition 2.12.)

Exercise 2.12. If s is a right special factor of $\boldsymbol{t}$, then $\bar{s}$ is also a right special factor of $\boldsymbol{t}$. Prove this also holds for left special factors of $\boldsymbol{t}$. (*Hint*: Use Exercise 1.7.)

Exercise 2.13. Prove that if s is a right special factor of $\boldsymbol{t}$, then $\mu(s)$ is a right special factor of $\boldsymbol{t}$. Prove this also holds for left special factors.

Exercise 2.14. Find another proof of Corollary 2.16 using Proposition 1.6, Exercise 1.8 and the previous exercise.

5. Formal languages

The goal of this section is to give a brief introduction to an aspect of formal language theory that involves the generation of languages.

A **language** over an alphabet A is a set of words over A. Several languages have a natural method for their generation. This has lead to the **Chomsky hierarchy** of classes of languages, with each class of languages in the hierarchy strictly containing the previous one. These are the *regular* languages, the *context-free* languages, the *context-sensitive* languages and the *recursively enumerable* languages. In what follows we will briefly illustrate the theory of regular and context-free languages using languages constructed from the Thue – Morse word. In particular, we prove that the language of factors of the Thue – Morse word is neither regular nor context-free, and the language of binary words that are not prefixes of the Thue – Morse word is context-free, but not regular. We also consider the language of binary words that are not factors of the Thue – Morse word.

5.1. Regular languages. Recall from Definition 2.1 that a finite deterministic automaton $\mathcal{A}$ over an alphabet A consists of a finite set of states Q, an initial state q_0, a set of final states F and a next state function, denoted by $\cdot : Q \times A \to Q$, that extends to $Q \times A^*$ multiplicatively.

A word $u \in A^*$ is **accepted** by the automaton if the state $q_0 \cdot u$ is a final state and **rejected** otherwise. For example, $abbbaaab$ is accepted by the automaton in Figure 2 while $abbbaaaba$ is rejected. The language of words accepted by an automaton $\mathcal{A}$ is denoted by $L(\mathcal{A})$.

$$L(\mathcal{A}) = \{w \in A^* : q_0 \cdot w \in F\}.$$

Definition 2.17. A language $L \subseteq A^*$ is **regular** if there exists a finite deterministic automaton $\mathcal{A}$ over A such that $L = L(\mathcal{A})$.

Let $\mathcal{A} = \langle A, Q, q_0, F, \cdot \rangle$ be a finite deterministic automaton and L the regular language accepted by $\mathcal{A}$. If $a_1, a_2, \ldots, a_p \in A$ with $a_1 a_2 \cdots a_p \in L$, then q_0, $q_0 \cdot a_1$, $q_0 \cdot (a_1 a_2)$, ..., $q_0 \cdot (a_1 \cdots a_p)$ describes a sequence of states of the automaton. If two of these states are the same (e.g., if $p \geq |Q|$), then there exist integers $0 \leq i < j \leq p$ such that $q_0 \cdot (a_1 \cdots a_i) = q_0 \cdot (a_1 \cdots a_j)$. So, for all $n \in \mathbb{N}$,

$$q_0 \cdot (a_1 \cdots a_i)(a_{i+1} \cdots a_j)^n = q_0 \cdot (a_1 \cdots a_i).$$

In particular, $q_0 \cdot (a_1 \cdots a_i)(a_{i+1} \cdots a_j)^n (a_{j+1} \cdots a_p)$ is a final state for all $n \in \mathbb{N}$. This observation is known as the *pumping lemma* for regular languages.

Lemma 2.18 (Pumping lemma for regular languages). *Suppose $L \subseteq A^*$ is a regular language. There exists an integer $p \geq 1$ such that for every word $w \in L$ with $|w| \geq p$, there is a factorization $w = (x, y, z)$ in A^* satisfying $y \neq \epsilon$, $|xy| \leq p$ and $xy^n z \in L$ for all $n \in \mathbb{N}$.*

The integer p in the statement of the lemma is called the **pumping length** of L. The terminology reflects the observation that a word $w \in L$ can be "pumped up" by repeating y an arbitrary number of times. The primary use of the pumping lemma is to prove that languages are not regular.

Proposition 2.19. *The set of factors of the Thue – Morse word $\boldsymbol{t}$ and the set of prefixes of $\boldsymbol{t}$ are not regular languages.*

PROOF. Let L denote the set of factors (respectively, prefixes) of the Thue – Morse word $\boldsymbol{t}$ and suppose that L is a regular language. Let p denote the pumping length of L and let w denote a factor (prefix) of $\boldsymbol{t}$ of length at least p. The pumping lemma for regular languages implies that there is a factorization $w = xyz$ such that $y \neq \epsilon$ and $xy^n z \in L$ for all $n \in \mathbb{N}$. Thus $xy^3 z$ is a factor of $\boldsymbol{t}$ for some nonempty word y, which contradicts the fact that $\boldsymbol{t}$ is overlap-free (Theorem 2.6). $\square$

The fact that the set of prefixes of $\boldsymbol{t}$ is not a regular language also follows from Exercise 2.18, which characterizes the infinite words whose prefixes form a regular language as those words that are ultimately periodic. (An infinite word w is **ultimately periodic** if there exist finite words x and y such that $w = xyyy\cdots$.)

It happens that regular languages are closed under complementation (Exercise 2.16), thus the following result is immediate from Proposition 2.19.

Corollary 2.20. *The set of binary words that are not factors of the Thue–Morse word* $\mathbf{t}$ *is not a regular language, nor is the language consisting of the binary words that are not prefixes of* $\mathbf{t}$.

Below we will see that one of these languages is context-free while the same question for the other language remains open.

Exercise 2.15. Let $\mathcal{A}$ denote the automaton defined by $A = \{a, b\}$, $Q = \{1, 2, 3, 4\}$, $q_0 = 1$, $F = \{4\}$ and next state function given by the following table.

$\cdot$	1	2	3	4
a	2	2	4	2
b	1	3	1	3

Draw the graph of $\mathcal{A}$ and describe the language $L(\mathcal{A})$ accepted by $\mathcal{A}$.

Exercise 2.16. The complement of a regular language is a regular language.

Exercise 2.17. Let $L = \{w \in \{0,1\}^* : |w|_0 = |w|_1\}$. Show that L is not a regular language.

Exercise 2.18. Let $\boldsymbol{w}$ be an infinite word over an alphabet A and let $L \subseteq A^*$ be the language consisting of the prefixes of $\boldsymbol{w}$. Then L is regular if and only if there exist finite words x and y such that $w = xyyy\cdots$. Conclude that the language of prefixes of the Thue–Morse word is not regular.

5.2. Context-free languages. Informally, a grammar provides a set of recursive rules for rewriting words over an alphabet A as words over a subset T of A. We will be concerned with the so-called *context-free grammars*.

Definition 2.21. A **context-free grammar** $\mathcal{G} = \langle V, T, P\rangle$ consists of an alphabet V of **variables**, an alphabet T of **terminal letters**, which is disjoint from V, and a finite set $P \subseteq V \times (V \cup T)^*$ of **productions**.

Suppose (v, u) is a production in a context-free grammar $\mathcal{G}$. The terminology "context-free" comes from the fact that v can be replaced by u regardless of the context in which v occurs. So if $w = xvy$ is a word in $(V \cup T)^*$, then the application of the rule (v, u) *produces* the new word $w' = xuy$. Often, letters from V will still occur in the word u of a production (v, u), so other productions can be used to replace these letters.

If (v, u) is a production and $w = xvy$ and $w' = xuy$, then we write $w \to w'$. Note that $v \to u$ for $(v, u) \in P$. More generally, given a sequence $w_0, \ldots, w_n$ of words over $V \cup T$ such that $w_0 \to w_1$, $w_1 \to w_2$, $\ldots$, $w_{n-1} \to w_n$, we write $w_0 \to w_1 \to \cdots \to w_n$ or $w_0 \xrightarrow{*} w_n$. Such a sequence is called a **derivation** from w_0 to w_n of length n, and we say that w_n is **derived** from w_0.

A context-free grammar $\mathcal{G} = \langle V, T, P\rangle$ generates a language $L(\mathcal{G}, v)$ by considering all the words over T that can be derived from a particular variable $v \in V$. Such languages are the *context-free languages*.

Definition 2.22. A language $L \subseteq T^*$ is a **context-free language** if there exists a context-free grammar $\mathcal{G} = \langle V, T, P\rangle$ and a variable $v \in V$ such that

$$L = L(\mathcal{G}, v) = \{w \in T^* : v \xrightarrow{*} w\}.$$

It happens that the class of context-free languages coincides with the class of languages accepted by *pushdown automata*, which will not be defined here. As in the case of regular languages, there exists a pumping lemma for context-free languages (see Exercise 2.22), whose primary use is to prove that a language is not context-free. In light of this, the proof of Proposition 2.19 also proves the following.

Proposition 2.23. *The set of factors of the Thue–Morse word and the set of prefixes of the Thue–Morse word are not context-free languages.*

Unlike for regular languages, context-free languages are not closed under complementation. Therefore, one can ask whether the complements of the languages of the above proposition form context-free languages.

Theorem 2.24. *The set of binary words that are not prefixes of the Thue–Morse word* $\boldsymbol{t}$ *is a context-free language.*

PROOF. If $w = w_0 w_1 \cdots w_r$ is a prefix of $\boldsymbol{t}$, then $w_0 = 0$, $w_{2n} = w_n$ and $w_{2m+1} = \overline{w}_m$ for all $n, m \in \mathbb{N}$ with $2n \leq |w|$ and $2m+1 \leq |w|$. Consequently, a word w is *not* a prefix of $\boldsymbol{t}$ if and only if w begins with 1, or $w = xay\bar{a}z$ with $|y| = |x| - 1$ and $a \in \{0,1\}$, or $w = xayaz$ with $|x| = |y|$ and $a \in \{0,1\}$. Since the class of context-free languages is closed under finite union (Exercise 2.21), we need only provide context-free grammars that generate each of these three languages. We call the languages A, B and C below. To simplify notation, we write $v \to \{u_1, u_2, \ldots, u_n\}$ to denote the set of productions $v \to u_1$, $v \to u_2$, $\ldots$, $v \to u_n$.

Consider the context-free grammar $\mathcal{A}$ with variables $\boldsymbol{\alpha}$ and $\boldsymbol{\beta}$, terminals $\{0,1\}$, and productions $\boldsymbol{\alpha} \to 1\boldsymbol{\beta}$ and $\boldsymbol{\beta} \to \{\epsilon, 0\boldsymbol{\beta}, 1\boldsymbol{\beta}\}$. Beginning with $\boldsymbol{\alpha}$, these productions generate all binary words beginning with 1: if w is a binary word beginning with 1, then

$$\boldsymbol{\alpha} \to 1\boldsymbol{\beta} \to 1(w_1\boldsymbol{\beta}) \to 1w_1(w_2\boldsymbol{\beta}) \to 1w_1w_2(w_3\boldsymbol{\beta}) \to \cdots \to w;$$

and every word in $L(\mathcal{A}, \boldsymbol{\alpha})$ begins with 1. Hence, $A = L(\mathcal{A}, \boldsymbol{\alpha})$ is context-free.

Next consider the grammar $\mathcal{B}$ with variables $\{\boldsymbol{\alpha}, \boldsymbol{\beta}, \boldsymbol{\gamma}\}$, terminals $\{0,1\}$ and productions

$$\boldsymbol{\alpha} \to \boldsymbol{\gamma}1\boldsymbol{\beta}, \quad \boldsymbol{\beta} \to \{\epsilon, 0\boldsymbol{\beta}, 1\boldsymbol{\beta}\}, \quad \boldsymbol{\gamma} \to \{0\boldsymbol{\gamma}0, 0\boldsymbol{\gamma}1, 1\boldsymbol{\gamma}0, 1\boldsymbol{\gamma}1, 00, 10\}.$$

We will show that $L(\mathcal{B}, \boldsymbol{\alpha})$ is the language of binary words $x0y1z$, where $x, y, z \in \{0,1\}^*$ with $|y| = |x| - 1$. Suppose $w = x'(a0)y1z$ with $a \in \{0,1\}$ and $|y| = |x'|$. By arguing as in the previous paragraph, we can show that $1\boldsymbol{\beta} \xrightarrow{*} 1z$ for any $z \in \{0,1\}^*$. Therefore, $\boldsymbol{\alpha} \to \boldsymbol{\gamma}1\boldsymbol{\beta} \xrightarrow{*} \boldsymbol{\gamma}1z$. So if $l = |y| - 1$,

$$\boldsymbol{\alpha} \xrightarrow{*} \boldsymbol{\gamma}1z \to (x'_0\boldsymbol{\gamma}y_l)1z \to x'_0(x'_1\boldsymbol{\gamma}y_{l-1})y_l1z \to \cdots \to x'\boldsymbol{\gamma}y1z \to w.$$

Thus $w \in L(\mathcal{B}, \boldsymbol{\alpha})$. The reverse containment is straightforward to prove. Similarly, $\{x1y0z : x, y, z \in \{0,1\}^*, |y| = |x| - 1\}$ is a context-free language, so B is as well.

Finally, consider the grammar $\mathcal{C}$ with variables $\{\boldsymbol{\alpha}, \boldsymbol{\beta}, \boldsymbol{\xi}\}$, terminals $\{0,1\}$ and productions $\boldsymbol{\alpha} \to \boldsymbol{\xi}0\boldsymbol{\beta}$, $\boldsymbol{\beta} \to \{\epsilon, 0\boldsymbol{\beta}, 1\boldsymbol{\beta}\}$ and $\boldsymbol{\xi} \to \{0\boldsymbol{\xi}0, 0\boldsymbol{\xi}1, 1\boldsymbol{\xi}0, 1\boldsymbol{\xi}1, 0\}$. Arguing as in the previous paragraph, it follows that $L(\mathcal{C}, \boldsymbol{\alpha})$ is the context-free language of all binary words $x0y0z$, where $x, y, z \in \{0,1\}^*$ and $|x| = |y|$. Similarly, $\{x1y1z : x, y, z \in \{0,1\}^*, |x| = |y|\}$ is a context-free language and so is C. □

So the complement of the prefixes of the Thue–Morse word $\boldsymbol{t}$ is a context-free language. The analogous question for all factors of $\boldsymbol{t}$ remains open.

Open Question. *Is the set of binary words that are not factors of the Thue–Morse word a context-free language?*

Towards answering this question, Narad Rampersad has recently shown that the language is not an *unambiguous* context-free language [Ram2007]. We outline the argument below.

There are, in a typical grammar, many ways to derive a word from a variable, as is illustrated in the following example.

Example. Consider the context-free grammar $\mathcal{G}$ with variable A, terminal a, and productions $A \to AA$ and $A \to a$. There are two distinct paths to aa:

$$A \to AA \to aA \to aa,$$
$$A \to AA \to Aa \to aa.$$

By taking the convention to always apply a production to the leftmost remaining variable, we obtained the so-called **leftmost derivations**. As the reader might imagine, this need not remove all the ambiguity.

Example. In the previous example there is exactly one leftmost derivation of aa. However, there are two leftmost derivations of aaa:

$$A \to AA \to aA \to aAA \to aaA \to aaa,$$
$$A \to AA \to AAA \to aAA \to aaA \to aaa.$$

Definition 2.25. A context-free language L is **unambiguous** if there exists a context-free grammar $\mathcal{G}$ generating L such that every $w \in L$ has exactly one leftmost derivation in $\mathcal{G}$.

Example. Let L be the context-free language generated by the context-free grammar $\mathcal{G}$ of the previous two examples. Then $L = \{a^n : n \geq 1\}$. This language is unambiguous because it can be generated by the context-free grammar with variable A, terminal a, and productions $A \to Aa$ and $A \to a$.

The following result of Noam Chomsky and Marcel-Paul Schützenberger is useful in proving that a given language is not unambiguous context-free.

Proposition 2.26 (Chomsky, Schützenberger [CS1963]**).** *If $L \subseteq A^*$ is an unambiguous context-free language, then the generating series $F_L(x) = \sum_{n\geq 0} |L \cap A^n| x^n$ is algebraic over $\mathbb{Q}(x)$.*

See Section 2 for the necessary notions regarding generating series. We apply this result with L equal to the language of binary words that are not factors of the Thue–Morse word $\boldsymbol{t}$. So $L \cap \{0,1\}^n$ is the set of binary words of length n that are not factors of $\boldsymbol{t}$. If $c_{\boldsymbol{t}}(n)$ is the number of factors of $\boldsymbol{t}$ of length n, then $|L \cap \{0,1\}^n| = 2^n - c_{\boldsymbol{t}}(n)$. It follows that the series $F_L(x)$ is algebraic if and only if the series $C_{\boldsymbol{t}}(x) = \sum_{n\geq 0} c_{\boldsymbol{t}}(n) x^n$ is algebraic. And $C_{\boldsymbol{t}}(x)$ is algebraic if and only if $S_{\boldsymbol{t}}(x) = \sum_{n\geq 0} s_{\boldsymbol{t}}(n) x^n$ is algebraic, where $s_{\boldsymbol{t}}(n) = c_{\boldsymbol{t}}(n+1) - c_{\boldsymbol{t}}(n)$ for all $n \in \mathbb{N}$.

Lemma 2.27 (Carlson [Car1921]**).** *A power series with integer coefficients and radius of convergence 1 is either rational or transcendental.*

We know from Section 4 that the sequence $\big(s_{\mathbf{t}}(n)\big)_{n\geq 0}$ is bounded between 2 and 4, so the series $S_{\mathbf{t}}(x)$ is either rational or transcendental by the above lemma. If the series $S_{\mathbf{t}}(x)$ is rational, then the sequence $s_{\mathbf{t}}(n)$ is ultimately periodic (Exercise 2.24). But this is not possible by (2.13). So the series $S_{\mathbf{t}}(x)$ is not algebraic and L is not unambiguous context-free.

Theorem 2.28 (Rampersad [Ram2007]). *The set of binary words that are not factors of the Thue–Morse word* $\mathbf{t}$ *is not unambiguous context-free.*

Exercise 2.19. Let $L = \{0^n1^n : n \in \mathbb{N}\}$. Show that L is a context-free language, but not a regular language.

Exercise 2.20. Define a context-free grammar $\mathcal{G} = \langle\{\alpha,\beta,\gamma,\delta\},\{0,1\},P\rangle$ with productions P given by

$$\begin{array}{lllll} \alpha \to \beta, & \alpha \to \gamma, & \beta \to \delta 0 \beta, & \beta \to \delta 0 \delta, & \\ \gamma \to \delta 1 \gamma, & \gamma \to \delta 1 \delta, & \delta \to 0 \delta 1 \delta, & \delta \to 1 \delta 0 \delta, & \delta \to \epsilon. \end{array}$$

Show that $L(\mathcal{G},\alpha)$ is the language of binary words with a different number of occurrences of 0s and 1s.

Exercise 2.21. If L and L' are context-free languages, then $L \cup L'$ is a context-free language.

Exercise 2.22 (Pumping Lemma for Context-Free Languages). Let L be a context-free language. There exists $p \in \mathbb{N}$ such that if $w \in L$ and $|w| \geq p$, then there exists a factorization $w = (u,v,x,y,z)$ satisfying $|v|,|y| > 0$, $|vxy| \leq p$, and $uv^ixy^iz \in L$ for each $i \geq 0$. (*Hint*: Argue that if $|w| \geq b^{|V|+1}$, where b is the maximum number of variables in the right-hand side of a production, then there is a derivation of the form $\xi \to v\xi y$ with $v, y \neq \epsilon$.)

Exercise 2.23. Prove that the language $L = \{a^nb^nc^n : n \in \mathbb{N}\} \subseteq \{a,b,c\}^*$ is neither regular nor context-free.

Exercise 2.24. If $\{a_0, a_1, a_2, \dots\}$ is a sequence in $\mathbb{R}$ taking only finitely many values and satisfying a linear recurrence relation (i.e.,

$$a_n = \gamma_1 a_{n-1} + \cdots + \gamma_k a_{n-k} \quad (\forall\, n \gg 0)$$

for fixed $\gamma_i \in \mathbb{R}$ and $k \in \mathbb{N}$), then $\{a_0, a_1, a_2, \dots\}$ is ultimately periodic.

6. The Tower of Hanoi

In the following we will use the Thue–Morse word to construct a solution to the Tower of Hanoi puzzle. Our exposition is based on an article by Jean-Paul Allouche, Dan Astoorian, Jim Randall and Jeffrey Shallit [AARS1994].

The **Tower of Hanoi** is a puzzle that appears to have been invented by the French number theorist François Édouard Anatole Lucas (1842–1891) under the pseudonym "N. Claus (of Siam)." It consists of a fixed number of disks, no two of which have the same radius, placed on top of each other in order of size with the largest disk on the bottom. See Figure 6. There are two other piles, which initially contain no disks. The goal of the puzzle is to move all the disks to one of the other piles according to the following rule: exactly one disk can be moved from one pile to another as long as the disk will not cover a smaller disk.

The Tower of Hanoi puzzle may be modelled by the directed graph in Figure 7. The three nodes each represent one of the piles, and the arrows represent moving a disk from one pile to another. A word over the alphabet $\{a, b, c, \bar{a}, \bar{b}, \bar{c}\}$ encodes a sequence of disk movements. For example, $a\bar{c}b$ encodes the following sequence of disk movements: move a disk from Pile 1 onto Pile 2; move a disk from Pile 1 onto Pile 3; move a disk from Pile 2 onto Pile 3. A solution to the problem of moving n disks from Pile i to Pile j amounts to constructing a word $\mathrm{Han}(n, i, j)$ over $\{a, b, c, \bar{a}, \bar{b}, \bar{c}\}$. We do this recursively in n.

If $n = 0$, then there are no disks to move, so the empty word provides a solution. Thus, let

$$\mathrm{Han}(0, i, j) = \epsilon.$$

The solution is nearly as simple for $n = 1$: a single letter chosen from $\{a, b, c, \bar{a}, \bar{b}, \bar{c}\}$ depending on the particular values of i and j.

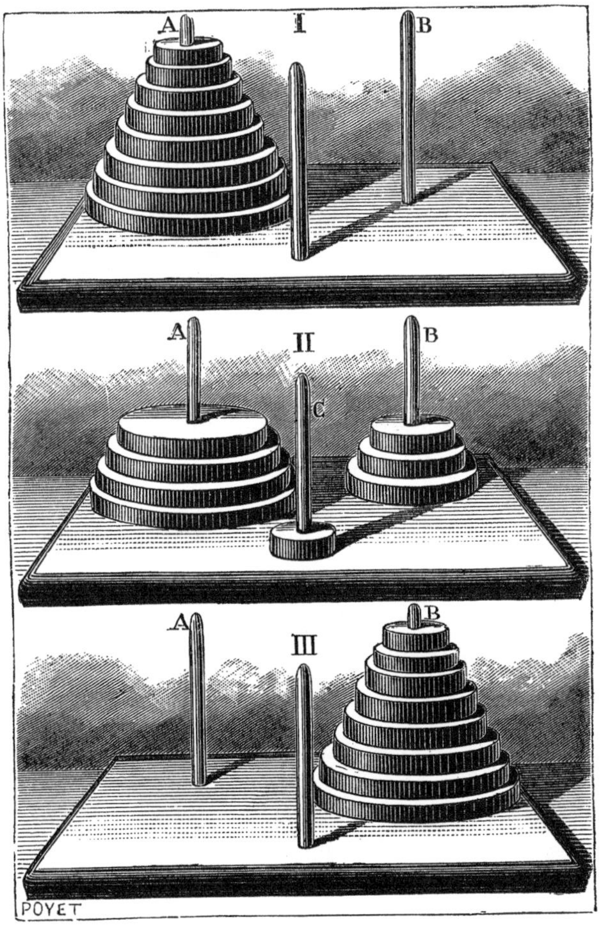

FIGURE 6. The Tower of Hanoi puzzle, reprinted with permission from LUCAS Edouard, *Récréations mathématiques,* Ed. Albert Blanchard, Paris [Luc1893].

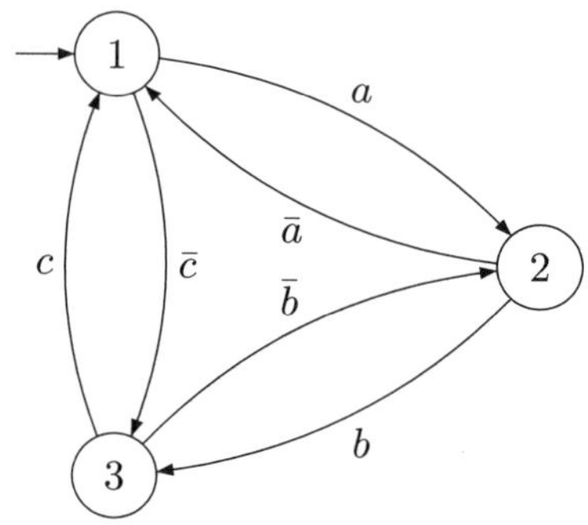

FIGURE 7. A model for the Tower of Hanoi puzzle.

Otherwise, suppose the solution $\mathrm{Han}(n-1, k, \ell)$ has been constructed for all $\{k, \ell\} \subseteq \{1, 2, 3\}$. To move n disks from Pile i to Pile j, we may move the top $n-1$ disks to an intermediate pile k, move the remaining disk from i to j, then move the $n-1$ disks from k to j. That is, we define

$$\mathrm{Han}(n, i, j) = \mathrm{Han}(n-1, i, k)\,\mathrm{Han}(1, i, j)\,\mathrm{Han}(n-1, k, j).$$

Examples. Referring to Figure 7, one sees that

$$\begin{aligned}
\mathrm{Han}(1,1,2) &= a, \\
\mathrm{Han}(2,1,3) &= \mathrm{Han}(1,1,2)\,\mathrm{Han}(1,1,3)\,\mathrm{Han}(1,2,3) \\
&= a\bar{c}b, \\
\mathrm{Han}(3,1,2) &= \mathrm{Han}(2,1,3)\,\mathrm{Han}(1,1,2)\,\mathrm{Han}(2,3,2) \\
&= (a\bar{c}b)(a)\,\mathrm{Han}(2,3,2) \\
&= (a\bar{c}b)(a)\,\mathrm{Han}(1,3,1)\,\mathrm{Han}(1,3,2)\,\mathrm{Han}(1,1,2) \\
&= (a\bar{c}b)(a)(c\bar{b}a).
\end{aligned}$$

Remark. This solution is optimal in the sense that it constructs a word of minimal length that solves the Tower of Hanoi puzzle. See Exercise 2.25.

In the above examples, $\mathrm{Han}(1,1,2)$ is a prefix of $\mathrm{Han}(2,1,3)$ and $\mathrm{Han}(2,1,3)$ is a prefix of $\mathrm{Han}(3,1,2)$. This generalizes as follows: for n even, $\mathrm{Han}(n,1,3)$ is a prefix of $\mathrm{Han}(n+1,1,2)$; for n odd, $\mathrm{Han}(n,1,2)$ is a prefix of $\mathrm{Han}(n+1,1,3)$. The limit of this sequence is the *Hanoi word*.

Definition 2.29. The **Hanoi word $\boldsymbol{h}$** is $\lim_{n\to\infty} H_n$, where

$$H_n = \begin{cases} \mathrm{Han}(n,1,3), & \text{if } n \text{ is even,} \\ \mathrm{Han}(n,1,2), & \text{if } n \text{ is odd.} \end{cases}$$

Example. From the previous example: $H_1 = a$, $H_2 = a\bar{c}b$ and $H_3 = a\bar{c}bac\bar{b}a$. Here are the first forty letters of the Hanoi word,

$$\boldsymbol{h} = a\bar{c}bac\bar{b}a\bar{c}b\bar{a}cba\bar{c}bac\bar{b}acb\bar{a}c\bar{b}a\bar{c}bac\bar{b}a\bar{c}b\bar{a}cba\bar{c}b\bar{a}\cdots.$$

Note that by cyclically permuting the letters a, b, c and $\bar{a}, \bar{b}, \bar{c}$ simultaneously, the word $\mathrm{Han}(n,1,2)$ becomes $\mathrm{Han}(n,2,3)$. This observation allows for a recursive construction of the words H_n.

Lemma 2.30. *For all $n \geq 1$,*

$$H_n = \begin{cases} H_{n-1}\,\bar{c}\,\sigma(H_{n-1}), & \text{for } n \text{ even,} \\ H_{n-1}\,a\,\sigma^2(H_{n-1}), & \text{for } n \text{ odd,} \end{cases}$$

where σ is the permutation of $\{a, b, c, \bar{a}, \bar{b}, \bar{c}\}$ defined by

$$\sigma(a) = b, \quad \sigma(b) = c, \quad \sigma(c) = a, \quad \sigma(\bar{a}) = \bar{b}, \quad \sigma(\bar{b}) = \bar{c}, \quad \sigma(\bar{c}) = \bar{a}.$$

To study the structure of the Hanoi word $\boldsymbol{h}$, it is useful to introduce two other infinite words $\boldsymbol{g}$ and $\boldsymbol{b}$ that encode the structure of $\boldsymbol{h}$. The word $\boldsymbol{g}$ is obtained from $\boldsymbol{h}$ by removing the bars from the barred letters of $\boldsymbol{h}$, and $\boldsymbol{b}$ is the binary word that records the location in $\boldsymbol{h}$ of the barred letters. We make this explicit.

For each $n \in \mathbb{N}$, let G_n denote the finite word over the alphabet $\{a, b, c\}$ that is the image of H_n under the morphism defined by $x \mapsto x$ and $\bar{x} \mapsto x$ for $x \in \{a, b, c\}$, and let B_n denote the binary word that is the image of H_n under the morphism defined by $x \mapsto 0$ and $\bar{x} \mapsto 1$ for $x \in \{a, b, c\}$. Let $\boldsymbol{g} = \lim_{n\to\infty} G_n$ and let $\boldsymbol{b} = \lim_{n\to\infty} B_n$.

Example. The table below lists the first four words of the sequences $(H_n)_{n\geq 1}$, $(G_n)_{n\geq 1}$ and $(B_n)_{n\geq 1}$.

n	1	2	3	4
H_n	a	$a\bar{c}b$	$a\bar{c}b\;ac\bar{b}\;a$	$a\bar{c}b\;ac\bar{b}\;a\bar{c}b\;\bar{a}cb\;a\bar{c}b$
G_n	a	acb	$acb\;acb\;a$	$acb\;acb\;acb\;acb\;acb$
B_n	0	010	010 001 0	010 001 010 100 010

The table suggests that $\boldsymbol{g} = \lim_{n\to\infty} G_n$ has a rather simple structure.

Proposition 2.31. *The word $\boldsymbol{g}$ is the periodic word $(acb)^\infty$.*

PROOF. From Lemma 2.30 we derive the following identity for G_n $(n \geq 1)$.

$$G_n = \begin{cases} G_{n-1}\,c\,\sigma(G_{n-1}), & \text{if } n \text{ is even,} \\ G_{n-1}\,a\,\sigma^2(G_{n-1}), & \text{if } n \text{ is odd.} \end{cases}$$

Induction on n establishes that G_n is of the form $(acb)^i a$ for n odd and of the form $(acb)^j$ for n even. ☐

Next is a characterization of $\boldsymbol{b} = \lim_{n\to\infty} B_n$ using the endomorphism $\nu\colon \{0,1\}^* \to \{0,1\}^*$ defined by $\nu(0) = 01$ and $\nu(1) = 00$. This morphism is known as the **period-doubling morphism**.

Proposition 2.32. *The word $\boldsymbol{b}$ is $\nu^\infty(0)$. That is, $\boldsymbol{b}$ is the fixed point of ν beginning with 0.*

PROOF. From Lemma 2.30 we derive the following identity for B_n $(n \geq 1)$.

$$B_n = \begin{cases} B_{n-1}1B_{n-1}, & \text{if } n \text{ is even,} \\ B_{n-1}0B_{n-1}, & \text{if } n \text{ is odd.} \end{cases}$$

Define a sequence v_n by

$$v_n = \begin{cases} B_n0, & \text{for } n \text{ even,} \\ B_n1, & \text{for } n \text{ odd,} \end{cases}$$

so that $B_{n+1} = v_n B_n$ for all $n \geq 0$. Then v_n and v_{n+2} end in the same letter and it follows that $v_{n+2} = v_{n+1} v_n v_n$.

We also have $\nu^{n+2}(0) = \nu^{n+1}(0)\nu^{n+1}(1) = \nu^{n+1}(0)\nu^n(0)\nu^n(0)$ for all $n \geq 0$. We conclude that $v_n = \nu^n(0)$ for all $n \geq 0$ since they satisfy the same recurrence and the same initial condition ($v_0 = B_0 0 = \nu^0(0)$).

Finally, for $i \in \mathbb{N}$ fixed, choose $n \in \mathbb{N}$ such that $|B_n| > i$. Then b_i is the letter in position i of B_n, hence it is the letter in position i of $v_n = \nu^n(0)$. It follows that $\boldsymbol{b} = \lim_{n \to \infty} \nu^n(0)$. □

It is perhaps not surprising to learn that $\boldsymbol{h}$, $\boldsymbol{g}$ and $\boldsymbol{b}$ are k-automatic sequences (indeed, for $\boldsymbol{g}$ and $\boldsymbol{b}$ this follows from Theorem 2.4). We leave the verification to the exercises, but display an automaton for $\boldsymbol{b}$ in Figure 8. We conclude, as promised, with a link between the Hanoi and Thue–Morse words.

Since $\boldsymbol{b}$ is a fixed point of the morphism ν, it immediately follows that

$$b_{2n} = 0 \quad \text{and} \quad b_{2n+1} = \bar{b}_n \tag{2.33}$$

for all $n \geq 0$. Comparing the Thue–Morse word $\boldsymbol{t}$ and the word $\boldsymbol{b}$,

$$\begin{aligned} \boldsymbol{t} &= 011010011001011010010110011010011001011 0\cdots \\ \boldsymbol{b} &= 0100010101000100010001010100010101000101\cdots \end{aligned}$$

we make the following observation.

Proposition 2.34. *If $\boldsymbol{t}$ is the Thue–Morse word and $\boldsymbol{b} = \lim_{n\to\infty} B_n$, then*

$$b_n = \begin{cases} 1, & \text{if } t_{n+1} = t_n, \\ 0, & \text{otherwise.} \end{cases}$$

PROOF. We prove the equivalent statement that $b_n = (t_{n+1} + t_n + 1) \bmod 2$. Let $s_n = (t_{n+1} + t_n + 1) \bmod 2$ for all $n \geq 0$. Then working modulo 2,

$$\begin{aligned} s_0 &= t_1 + t_0 + 1 = 1 + 0 + 1 = 0, \\ s_{2n} &= t_{2n+1} + t_{2n} + 1 = \bar{t}_n + t_n + 1 = 0, \\ s_{2n+1} &= t_{2n+2} + t_{2n+1} + 1 = t_{n+1} + \bar{t}_n + 1 \\ &= t_{n+1} + t_n = s_n - 1 = \bar{s}_n. \end{aligned}$$

Comparing with (2.33), we see b_n and s_n satisfy the same recurrences. □

Since the Hanoi word $\boldsymbol{h}$ can be reconstructed from the words $\boldsymbol{g}$ and $\boldsymbol{b}$, it follows from Proposition 2.31 and Proposition 2.34 that $\boldsymbol{h}$ can be constructed directly from the Thue–Morse word.

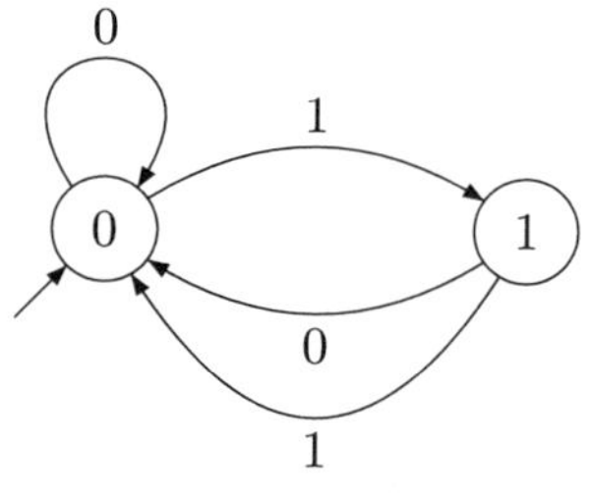

FIGURE 8. The automaton for $\boldsymbol{b}$.

Theorem 2.35. *The Hanoi word* $\boldsymbol{h}$ *is obtained from the Thue – Morse word* $\boldsymbol{t}$ *by placing a bar over the* n*th letter of* $(acb)^\infty$ *if and only if* $t_n = t_{n+1}$.

Exercise 2.25 ([AARS1994]). This aim of this exercise is to prove that the Hanoi word provides an optimal solution to the Tower of Hanoi problem.

(a) Let $T_n = |\operatorname{Han}(n, i, j)|$ for $i \neq j$ and $n \geq 0$. Argue that $T_n = 2T_{n-1} + 1$ for all $n \geq 0$.

(b) Show that $T_n = 2^n - 1$ for all $n \geq 0$.

(c) Prove that $\operatorname{Han}(n, i, j)$ provides an optimal solution to the Tower of Hanoi problem containing n disks. (*Hint*: Argue that any optimal solution requires at least T_n disk movements.)

Exercise 2.26. Let $\nu(0) = 01$ and $\nu(1) = 00$, and let $\boldsymbol{b} = \lim_{n\to\infty} \nu^n(0)$. Then $\boldsymbol{b}$ satisfies the following for all $n \geq 0$.

(a) $b_{4n+1} = 1$.
(b) $b_{4n+3} = b_n$.
(c) $b_n = 1$ if and only if the binary expansion of n ends with an odd number of 1s.

(*Hint*: To prove (c), show that the automaton in Figure 8 outputs $\boldsymbol{b}$. Alternatively, use Propositions 1.2 and 2.34.)

Exercise 2.27. The bar-free Hanoi word $\boldsymbol{g} = (acb)^\infty$ is 3-automatic.

Exercise 2.28 ([AARS1994]). The Hanoi word $\boldsymbol{h}$ is a fixed point of the morphism $\varphi\colon \{a, b, c, \bar{a}, \bar{b}, \bar{c}\}^* \to \{a, b, c, \bar{a}, \bar{b}, \bar{c}\}^*$ defined by

$$\varphi(a) = a\bar{c}, \qquad \varphi(b) = c\bar{b}, \qquad \varphi(c) = b\bar{a},$$
$$\varphi(\bar{a}) = ac, \qquad \varphi(\bar{b}) = cb, \qquad \varphi(\bar{c}) = ba.$$

Exercise 2.29. The Hanoi word $\boldsymbol{h}$ is a 2-automatic sequence. Find a finite deterministic automaton that outputs $\boldsymbol{h}$.

Exercise 2.30 ([AB1992, AAB+1995]). Let $\boldsymbol{s}$ denote the word obtained from the periodic word $(101\circ)^\infty = 101 \circ 101 \circ 101 \circ \cdots$ over the alphabet $\{0, 1, \circ\}$ by replacing the symbols $\circ$ with successive terms of the word $\boldsymbol{s}$. Thus $\boldsymbol{s}$ begins as

$$\boldsymbol{s} = 101\mathbf{1}101\mathbf{0}101\mathbf{1}101\mathbf{1}101\mathbf{1}101\mathbf{0}101\cdots.$$

Prove that $\boldsymbol{s} = \bar{\boldsymbol{b}}$. (*Hint*: Show that $s_n = 0$ if $t_n = t_{n+1}$ and $s_n = 1$ otherwise. Alternatively, note that, by construction, $s_{4n+3} = s_n$, $s_{4n} = 1$, $s_{2n+2} = 1$ and $s_{4n+1} = 0$ for all $n \geq 0$ and use Exercise 2.26.)
(This construction is an example of a *Toeplitz word* or a *Toeplitz sequence*. For more information see [AB1992].)

CHAPTER 3

Square-Free Words

A finite or infinite word m is called **kth-power-free** if there is no word $u \neq \epsilon$ such that u^k is a factor of m. The special case of **square-free** words has been studied at least since [Thu1906], where Thue used the notion in his study of patterns in infinite words (see Section 3 of Chapter 5). We indicate some elements of the theory below.

1. One example, three constructions

There are only 6 nonempty square-free words on two letters, namely

$$0, \quad 1, \quad 01, \quad 10, \quad 010, \quad 101.$$

By contrast, there are infinitely many square-free words on three letters. Indeed, even infinite ones (see Exercises 3.1 and 3.2). Below, we look at an infinite square-free word $\boldsymbol{m}$ on three letters related to the Thue–Morse word.

Construction I (Braunholtz [NBT1963]). Starting to the right of the initial 0 in the Thue–Morse word $\boldsymbol{t}$, record the lengths of blocks of 1s in $\boldsymbol{t}$:

$$\begin{array}{rl} \boldsymbol{t}\colon & 011\ 01\ 0\ 011\ 0\ 01\ 011\ 0\cdots \\ \boldsymbol{m}\colon & 2\quad 1\ \ 0\ 2\quad 0\ 1\quad 2\qquad \cdots \end{array}$$

The word $\boldsymbol{m}$ thus constructed is square-free. Indeed, suppose $u = a_1 \cdots a_n$ is a word such that uu is a factor of $\boldsymbol{m}$. Then

$$\underbrace{(0\,1^{a_1}\cdots 0\,1^{a_n})\,(0}\overbrace{1^{a_1}\cdots 0\,1^{a_n})0}$$

is a factor of $\boldsymbol{t}$, providing $\boldsymbol{t}$ with an overlap and contradicting Theorem 2.6.

Construction II (Thue [Thu1906]). It is clear from Theorem 2.6 that the Thue–Morse word $\boldsymbol{t}$ does not have 111 as a factor. This allows us to define $\boldsymbol{m}$ above as the (unique) preimage of $\boldsymbol{t}$ under the morphism $\gamma\colon (0,1,2) \mapsto (0,01,011)$.

Our final construction of $\boldsymbol{m}$ is based on two auxillary words $\boldsymbol{t}'$ and $\boldsymbol{t}''$ derived from $\boldsymbol{t}$.

Construction III (Morse, Hedlund [MH1944]). Construct the word $\boldsymbol{t}'$ by reading the letters of $\boldsymbol{t}$ two at a time and converting from base 2 to base 4: $(00, 01, 10, 11) \mapsto (0, 1, 2, 3)$. That is, $t'_n = 2t_n + t_{n+1}$.

$$\begin{array}{rl} \boldsymbol{t}\colon & 0\,1\,1\,0\,1\,0\,0\,1\,1\,0\,0\,1\,0\cdots \\ \boldsymbol{t}'\colon & 1\,3\,2\,1\,2\,0\,1\,3\,2\,0\,1\,2\ \ \cdots \end{array} \tag{3.1}$$

Next, construct $\boldsymbol{t}''$ by reducing the letters of $\boldsymbol{t}'$ modulo 3.

$$\begin{array}{rl} \boldsymbol{t}'\colon & 1\,3\,2\,1\,2\,0\,1\,3\,2\,0\,1\,2\cdots \\ \boldsymbol{t}''\colon & 1\,0\,2\,1\,2\,0\,1\,0\,2\,0\,1\,2\cdots \end{array}$$

Finally, let $\boldsymbol{t}''+1^\infty$ denote the word on $\{0,1,2\}$ defined by $(\boldsymbol{t}''+1^\infty)_n \equiv t''_n+1 \bmod 3$.

$$\begin{array}{rl} \boldsymbol{t} : & 0\,1\,1\,0\,1\,0\,0\,1\,1\,0\cdots \\ \boldsymbol{p} : & 0 \quad\; 3 \;\; 5\,6 \quad\; 9\cdots \\ \boldsymbol{m} : & 2 \quad\; 1 \;\; 0\,2 \end{array}$$

FIGURE 1. Defining $\boldsymbol{p}$ by $p_n = 2n + t_n$, the relationship between $\boldsymbol{t}$, $\boldsymbol{p}$ and $\boldsymbol{m}$ is uncovered.

Proposition 3.2. *The words* $\boldsymbol{m}$ *and* $\boldsymbol{t}'' + 1^{\infty}$ *coincide.*

PROOF. Note that $t''_n \equiv t_{n+1} - t_n \bmod 3$. We claim that this expression also equals $m_n - 1 \bmod 3$, from which the desired result follows.

To prove the claim, we consider the auxiliary word $\boldsymbol{p}$ defined by recording the position of the nth occurrence of 0 in $\boldsymbol{t}$. That is, $p_n = N$ if $t_N = 0$ and $|t_0t_1\cdots t_{N-1}|_0 = n - 1$:

$$\begin{array}{rl} \boldsymbol{t}: & 0110100110\cdots \\ \boldsymbol{p}: & 0 \quad\; 3 \;\; 56 \quad\; 9\cdots \end{array}$$

From the definition of the Thue–Morse word, we know that

$$p_n = \begin{cases} 2n, & \text{if } t_n = 0, \\ 2n+1, & \text{otherwise.} \end{cases} \tag{3.3}$$

In other words, $p_n = 2n + t_n$. Evidently, m_n equals 0, 1 or 2 according to whether $p_{n+1} - p_n - 1$ equals 0, 1 or 2 (see Figure 1). But this last expression is also equal to the value of $t_{n+1} - t_n + 1$, proving the claim. □

We conclude with alternative constructions of $\boldsymbol{t}'$ and $\boldsymbol{t}''$, in the spirit of Proposition 1.5.

Proposition 3.4 (Morse, Hedlund [MH1944]**).** *The endomorphism* α *of words over the alphabet* $\{0,1,2,3\}$ *defined by* $\alpha\colon (0,1,2,3) \mapsto (12,13,20,21)$ *satisfies* $\boldsymbol{t}' = \alpha^{\infty}(1)$.

PROOF. The proof rests on the following identity (see Exercise 3.5):

$$[\mu^n(0)1]' = \alpha^n(1) \quad \text{for all } n \geq 0,$$

where μ is the Thue–Morse morphism and $[\mu^n(0)1]'$ denotes the word constructed from $\mu^n(0)1$ in the same manner that $\boldsymbol{t}'$ is constructed from $\boldsymbol{t}$ (reading two letters at a time and converting from base 2 to base 4).

Example. Taking $n = 2$, we have $\mu^2(0)1 = 01101$, $[01101]' = 1321$ and $\alpha^2(1) = \alpha(13) = 1321$.

Continuing with the proof, we prove that $\mu^n(0)1$ is a prefix of $\boldsymbol{t}$ for all n. Indeed, we have seen that $\boldsymbol{t} = \mu^{\infty}(0)$ and moreover $\mu^{n+1}(0) = \mu^n(0)\mu^n(1)$, since μ is a morphism. Since $\mu^n(1)$ begins in 1, the result follows. Using the identity above, we conclude that $\alpha^n(1)$ is a prefix of $\boldsymbol{t}'$, finishing the proof. □

Let β be the endomorphism of $\{0,1,2,3\}^*$ defined by $\beta\colon (0,1,2,3) \mapsto (0,1,2,0)$. An immediate corollary of Proposition 3.4 is that $\boldsymbol{t}'' = \beta\bigl(\alpha^{\infty}(1)\bigr)$. We use this fact to prove a different characterization of $\boldsymbol{t}''$ due to Marshall Hall [Hal1964]. We call the endomorphism σ that he uses the **morphism of Hall** in what follows.

Proposition 3.5 (Hall [Hal1964]). *Let σ denote the morphism on $\{0,1,2\}^*$ defined by $\sigma\colon (0,1,2) \mapsto (12, 102, 0)$. Then $\boldsymbol{t}'' = \sigma^\infty(1)$.*

PROOF. The proof rests on the altogether not obvious identities of Exercise 3.6, which may be proved by parallel induction on n:

$$\sigma^{n+1}(0) = \beta\big(\alpha^n(12)\big),$$
$$\sigma^{n+1}(1) = \beta\big(\alpha^n(132)\big),$$
$$\sigma^{n+1}(2) = \beta\big(\alpha^n(0)\big).$$

The proof is now immediate, for we have

$$\begin{aligned}\sigma^{n+2}(1) &= \sigma^{n+1}(102) = \beta\alpha^n(132120)\\ &= \beta\alpha^{n+1}(132) = \beta\alpha^{n+2}(1)\beta\alpha^{n+1}(2).\end{aligned}$$

In particular, σ^n and $\beta\alpha^n$ have common prefixes of strictly increasing lengths for all $n \geq 2$. □

Remark. The Thue – Morse word and its derivatives are not the only infinite words the reader has seen that can be described as the fixed point of a morphism. See Exercise 3.7 for a Fibonacci example.

Exercise 3.1 ([Lot1997, Lemma 2.1.2]). Fix an alphabet A and let $\mathcal{P}$ be a property of elements of A^* which is closed under taking factors. Show that the following two statements are equivalent if and only if A is finite.

(a) The set $L_{\mathcal{P}}$ of words $w \in A^*$ having propery $\mathcal{P}$ is infinite.
(b) There exists an infinite word $\boldsymbol{w}$ on A whose (finite) factors all have property $\mathcal{P}$.

Exercise 3.2 ([MH1944]). For each infinite word $\boldsymbol{a} = a_0a_1\cdots$ on $A = \{a,b\}$, define an infinite word $\boldsymbol{b} = b_0b_1\cdots$ on $B = \{a,b,c\}$ by

$$b_n = \begin{cases} a, & \text{if } a_na_{n+1} \in \{aa, bb\},\\ b, & \text{if } a_na_{n+1} = ab,\\ c, & \text{if } a_na_{n+1} = ba.\end{cases}$$

Prove that if $\boldsymbol{a}$ is overlap-free, then $\boldsymbol{b}$ is square-free.

Exercise 3.3. Show that $\boldsymbol{t}'$ and $\boldsymbol{t}''$ are square-free. (*Hint*: Use Exercise 3.2 with $\boldsymbol{a} = \boldsymbol{t}$.)

Exercise 3.4 ([AARS1994]). Show that the Hanoi word $\boldsymbol{h}$ (Definition 2.29) is square-free.

Exercise 3.5. Complete the proof of Proposition 3.4. That is, verify the identity $\big[\mu^n(0)1\big]' = \alpha^n(1)$. (*Hint*: A proof by induction seems natural. You may want to simultaneously verify

$$\big[\mu^n(0)0\big]' = \alpha^n(0), \quad \big[\mu^n(1)0\big]' = \alpha^n(2), \quad \big[\mu^n(1)1\big]' = \alpha^n(3),$$

where given a word w, the word w' is built as in (3.1) by reading the letters of w two at a time and converting from binary.)

Exercise 3.6. Given α, β and σ as in Proposition 3.5, verify that the identities below hold for all n:

$$\sigma^{n+1}(0) = \beta\big(\alpha^n(12)\big), \quad \sigma^{n+1}(1) = \beta\big(\alpha^n(132)\big), \quad \sigma^{n+1}(2) = \beta\big(\alpha^n(0)\big).$$

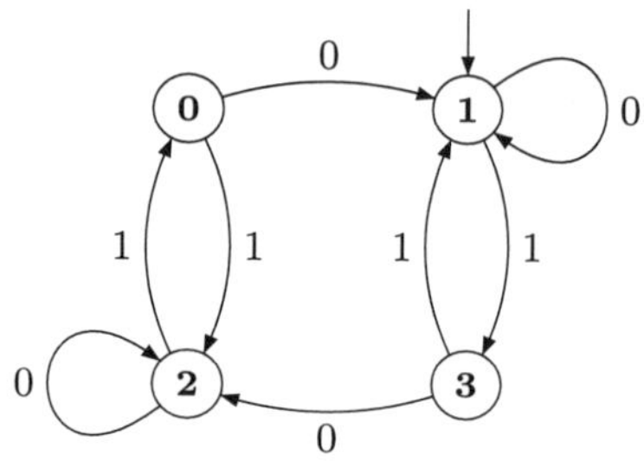

FIGURE 2. An automaton that outputs $\boldsymbol{t}'$.

Exercise 3.7. Let Φ denote the composition $\mathbf{E} \circ \mathbf{D}$ of Christoffel morphisms from Chapter 2 of Part I, i.e., $(x, y) \stackrel{\Phi}{\mapsto} (xy, x)$. Prove that $\Phi^\infty(x)$ is the Fibonacci word $\boldsymbol{f}$ defined in Exercise 1.5 of Part I.

Exercise 3.8. An automaton for $\boldsymbol{t}'$.

(a) Prove that the automaton in Figure 2 outputs the word $\boldsymbol{t}'$. (*Hint*: Use the argument in the proof of Theorem 2.4 and Proposition 3.4.)

(b) Develop a combinatorial characterization of t'_n based on the binary expansion of n. (*Hint*: By definition, $t'_n = 2t_n + t_{n+1}$, so t'_n is either 0 or 1 if $t_n = 0$ and is either 2 or 3 if $t_n = 1$.)

2. Square-free morphisms and codes

A morphism $h\colon A^* \to B^*$ is a **square-free morphism**, or more simply h is **square-free**, if for every square-free word w over A, the image $h(w)$ is square-free over B. Exercise 3.10 indicates a strong connection between square-free morphisms and the more general notion of kth-power-free morphisms.

Evidently, the trivial morphism $h\colon A^* \to \{\epsilon\}$ is square-free. We rule out this case in what follows; by "morphism" we shall always mean "nontrivial morphism."

A complete characterization of square-free morphisms does not exist, though Maxime Crochemore [Cro1983b] showed that the monoid of square-free morphisms is not finitely generated.[1] One evident fact is the following.

Proposition 3.6. *If $h\colon A^* \to B^*$ is a square-free morphism and if $h^\infty(a)$ exists for some $a \in A$, then $h^\infty(a)$ is a square-free word.*

Remark. The converse to this proposition is false, as verified by the morphism of Hall: we have seen that $\sigma^\infty(1) = \boldsymbol{t}''$ is square-free, and clearly 101 is square-free, but $\sigma(101) = 1\,0\,\underline{2\,1}\,\underline{2\,1}\,0\,2$ is not.

Short of demanding square-freeness, one might ask if h is at least *k-square-free*: a morphism h is called **k-square-free** if it preserves the square-free property of words of length at most k. This notion was introduced in developing criteria to test whether or not h is square-free. Early results in this direction include the following.

Proposition 3.7 (Crochemore [Cro1982]**).** *If $|A| = 3$, a morphism $h\colon A^* \to B^*$ is square-free if and only if h is 5-square-free.*

---

[1]More generally, the monoid of kth-power-free morphisms ($k \geq 3$) and overlap-free morphisms are also not finitely generated. (See [RW2002] and [Ric2003].)

This leads naturally to the notion of *test sets*: a set $T \subseteq A^*$ is a **test set** for the square-freeness of h if one may deduce that h is square-free by checking that $h(t)$ is square-free for all $t \in T$. The proposition states that a test set for "ternary morphisms" is the set of all square-free words on three letters of length at most five. Fortunately, each such word is a factor of a square-free word of length equal to five, so a minimal test set contains a quite manageable thirty elements. (See Exercise 3.9.)

Theorem 3.8 (Crochemore [Cro1982]**).** *A morphism $h\colon A^* \to B^*$ is square-free if and only if it is k-square-free for*

$$k = \max\left\{3, 1 + \left\lceil \frac{M(h) - 3}{m(h)} \right\rceil \right\},$$

where $\lceil\ \rceil$ is the ceiling function, $M(h) = \max\{|h(a)| : a \in A\}$ and $m(h) = \min\{|h(a)| : a \in A\}$.

Example. If h is a uniform morphism, then $M(h) = m(h)$ and $k = 3$. The theorem then reads, "3-*square-free uniform morphisms are square-free*," a statement which also follows from Theorem 3.11 below.

Crochemore's theorem gives an upper bound on the size of test sets in the general case. See also [HYY2003], where Hung-Kuei Hsiao, Yow-Tzong Yeh and Shyr-Shen Yu give a similar test set involving

$$\max\{k : h(a) \cap B^* h(A^k) B^* \neq \varnothing\}.$$

A more precise description of test sets, at least in the setting of kth-power-free morphisms ($k \geq 3$), has been undertaken by Gwénaël Richomme and Francis Wlazinski [RW2004, RW2007].

The balance of this chapter is devoted to two important k-square-free tests for a morphism to be square-free. We begin with some elementary properties of square-free morphisms.

Lemma 3.9. *Let $h\colon A^* \to B^*$ be a (nontrivial) morphism, and let C denote the set of images $\{h(a) : a \in A\}$. If h is square-free, then:*

(i) *h is nonerasing,*
(ii) *h is injective on its alphabet,*
(iii) *no element $c \in C$ is the prefix of another element $c' \in C$,*
(iv) *no element $c \in C$ is the suffix of another element $c' \in C$.*

PROOF. (i) Suppose that h is erasing (i.e., there is a letter $a \in A$ with $h(a) = \epsilon$). Since h is not the trivial morphism, there is some $b \in A$ such that $h(b) \neq \epsilon$. But then, bab is square-free while $h(bab) = h(b)h(b)$ is not square-free.

(iii) Suppose $a, b \in A$ and $x \in B^*$ are such that $h(b) = h(a)x$. Then $h(ab) = h(a)h(a)x$ fails to be square-free.

(ii) & (iv) Follow the proof of (iii). □

After the lemma, we may restrict our search for square-free morphisms to nonerasing injective morphisms h. If, moreover, h is 2-square-free, then the proof of the lemma realizes $h(A)$ as a *code* in B^*.

Definition 3.10. A **code** over an alphabet B is a set of words $C \subseteq B^*$ such that every $w \in C^*$ has a unique factorization $w = (c_1, c_2, \cdots, c_r)$ with $c_j \in C$.

In the context of codes, Properties (iii) and (iv) of the above lemma are the definitions of *prefix code* and *suffix code:* we say that a code C is a **prefix code** (respectively, **suffix code**) if no element of C is the prefix (respectively, suffix) of another element of C. A code C that is both a prefix code and a suffix code is called a **bifix code**. If, moreover, no element of C is a factor of another element of C, then C is called an **infix code**. An important class of infix codes are **uniform codes**, that is, codes C whose elements have a common length.

Examples. The (nonuniform) code $\{01,12,0210\}$ is infix while the code $\{0,101\}$ is not. The code $\{01, 02, 12\}$ is uniform, while $\{10, 01, 11, 00, 10011\}$ is not even a code.

As we will focus our attention on (nonerasing, injective) 2-square-free morphisms, we freely use the language of codes in what follows. Moreover, if $h\colon A^* \to B^*$ is a morphism giving rise to a code (that is, $C(h) := h(A)$ is a code in B^*), then we transport properties naturally defined for C to h and vice versa. For example, we speak freely below of "k-square-free codes" (a code coming from a k-square-free morphism) and "infix morphisms" (a morphism whose associated code is infix).

Exercise 3.9. A few square-free facts about ternary alphabets A:

(a) Every square-free word over A of length less than five appears as a prefix of some square-free word over A of length five.
(b) There are 30 square-free word of length five.
(c) There are square-free words in A^7 which cannot be extended to longer square-free words. Seven is the minimal integer with this property.

Exercise 3.10 ([BEM1979]). A nonerasing morphism $h\colon A^* \to B^*$ is said to be kth-power-free if $h(w)$ is kth-power-free for each kth-power-free word w. Show that if h is a square-free morphism such that

(a) $h(A)$ is infix, and
(b) if $|h(a)| > 1$ then $h(a)$ does not begin and end in the same letter,

then h is kth-power-free for all $k \geq 2$.

3. A 3-square-free test for square-freeness

Our first theorem goes back to Axel Thue's work. It has also been given in the paper by Dwight Bean, Andrzej Ehrenfeucht and George McNulty [BEM1979].

Theorem 3.11. *A 3-square-free infix morphism is square-free.*

In particular, infix morphisms have a test set composed of square-free words of length at most 3. Before starting the proof, we consider a notion closely related to infix.

Definition 3.12. A code C is **comma-free** if for all $ucv \in C^*$, with $c \in C$, one has $u, v \in C^*$.

Example. The code $\{0121, 01021, 20102\}$ is comma-free. (This is more easily seen after the forthcoming lemma.) The Thue–Morse morphism $\mu\colon (0,1) \mapsto (01,10)$ is not comma-free because $\mu(00) = 0101 = 0\mu(1)1$.

Lemma 3.13. *A comma-free code is infix. Conversely, a 2-square-free infix code is comma-free.*

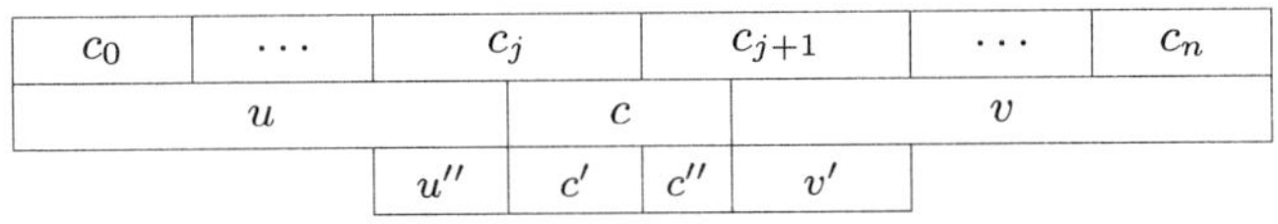

FIGURE 3. A factorization as ucv of a word $c_0 \cdots c_n$ in C^* for a 2-square-free infix code C.

PROOF. Let $C = C(h)$ be a comma-free code associated to some morphism $h : A^* \to B^*$. Assume that $ucv \in C$ for some $c \in C$. Since C is comma-free, this implies $u, v \in C^*$, and since C is a code, this forces $u = v = \epsilon$.

Conversely, suppose C is 2-square-free and infix. Consider a word $ucv \in C^*$ with $c \in C$ and $u, v \in B^*$. Then $ucv = c_0 \cdots c_n$ for unique $c_0, \ldots, c_n \in C$. First, we establish the factorization in Figure 3 for some $0 \leq j < n$ and $c', v' \neq \epsilon$.

The index j is chosen so that $|c_0 \cdots c_{j-1}| \leq |u| < |c_0 \cdots c_j|$. By the infix property of C, c cannot be a factor of c_j, so there are two cases: (i) c and c_j begin and end at the same point within ucv, or (ii) the prefix uc eclipses the right edge of c_j. In the latter case, again because C is infix, c_{j+1} cannot end before (or at) the end of c. So we have factorizations

$$c_j = u''c' \quad c = c'c'' \quad c_{j+1} = c''v'$$

for some words $u'', c', c'', v' \in B^*$, with $c', v' \neq \epsilon$.

Noting that $c_j c$ contains a square, we deduce by the 2-square-free property of C that $c_j = c$. The first case above then satisfies $u, v \in C^*$ and we are done.

In the remaining case (u'' and c'' are both nonempty), we see that cc_{j+1} also contains a square (forcing $c_j = c = c_{j+1}$). We further suppose that $|c'| > |c''|$ (assuming $|c'| \leq |c''|$ instead, one reaches the same conclusion). Since c' is a suffix of $c_j = c$, we see that c'' is a suffix of c'. Thus $c = c'c'' = xc''c''$ (for some $x \in B^*$) contains a square and C is not even 1-square-free. $\square$

PROOF OF THEOREM 3.11. Suppose $h\colon A^* \to B^*$ is a 3-square-free infix morphism and set $C = C(h)$. Assume the result is false and let $w = a_0 a_1 \cdots a_n$ ($a_i \in A$) be a shortest square-free word such that $h(w)$ contains a square uu ($u \in B^* \setminus \{\epsilon\}$). We have $n \geq 3$ by hypothesis.

Writing $h(a_i) = c_i$, we may assume that the product uu starts at the beginning of or within c_0 and ends within or at the end of c_n (otherwise we could have chosen a shorter word w). We claim that the factorization in Figure 4(a) cannot happen and that the true picture is Figure 4(b) for some $0 < j < n$.

Indeed, if the first picture holds true, one would have c_1 as a factor of u (since $n > 1$), and hence as a factor of c_0, violating the infix property of C. Turning to the second picture, we refine it by introducing factorizations $c_j = ps$, $c_0 = p's'$ and $c_n = p''s''$ as in Figure 5 (with $s', p, p'' \neq \epsilon$).

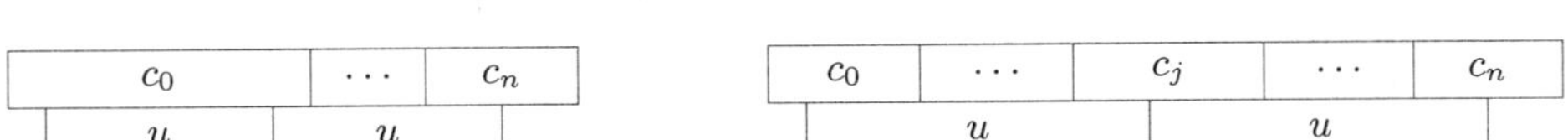

(a) u is a factor of c_0.

(b) c_j straddles the factorization uu.

FIGURE 4. Potential instances of the square uu within $h(w)$.

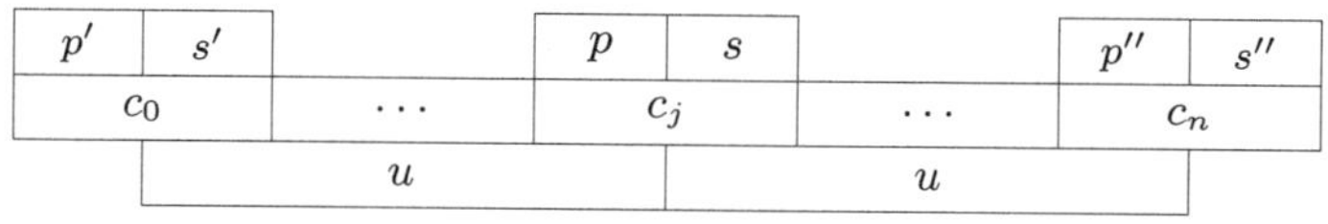

FIGURE 5. An instance of the square uu within $h(w)$.

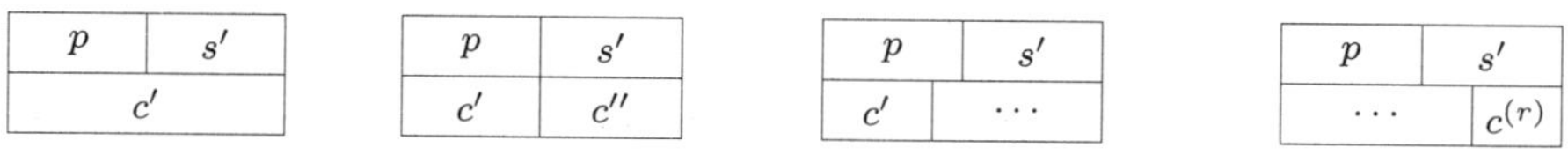

(a) $ps' \in C$. (b) $p, s' \in C$. (c) p begins in c'. (d) s' ends in $c^{(r)}$.

FIGURE 6. Possible factorizations of ps' inside C^*.

Note that either c_1 or c_{n-1} is a factor of u (since $n \geq 3$). Say we are in the first case (the other one being symmetric). Then

$$c_j \cdots c_n = pus'' = ps'c_1 \cdots c_{j-1}ps''.$$

Moreover, since C is a comma-free code by Lemma 3.13, the factorization $(ps')c_1(c_2 \cdots c_{j-1}ps'') \in C^*$ means that $ps' \in C^* \setminus \{\epsilon\}$. Writing $ps' = c'c'' \cdots c^{(r)}$, there are four cases to consider. See Figure 6.

Cases (c) and (d) of Figure 6 are excluded because C is bifix, that is, because c' is a prefix of c_j and $c^{(r)}$ is a suffix of c_0, respectively.

In Case (b), one has $p = c' = c_j$ and $s' = c'' = c_0$ (since C is bifix). Moreover, $p' = \epsilon$ and $s = \epsilon$. We now have the factorizations

$$u = c_0c_1 \cdots c_j = c_{j+1} \cdots c_{n-1}p'',$$

or $c_0c_1 \cdots c_js'' = c_{j+1} \cdots c_n$. The prefix property of C then gives us that $s'' = \epsilon$ and $p'' = c_n$. Indeed, c_0 is a prefix of c_{j+1} (or vice versa) and hence they are equal. Continuing this line of reasoning, peeling off the left-most c_i at each step, we are left with one of three possibilities depending on whether $n - j$ is less than, equal to or greater than $j + 1$:

$$c_\ell \cdots c_js'' = \epsilon \qquad s'' = \epsilon \qquad \epsilon = c_\ell \cdots c_{n-1}p''.$$

Since h is nonerasing, the only allowable result above is the middle one. So we conclude that $p'' = c_n$, $n = 2j+1$ and $c_k = c_{j+1+k}$ for $0 \leq k \leq j$, i.e., w is a square.

In Case (a) of Figure 6, we have $c' = ps' = c_j$ (C is prefix) and $s = s'$. It follows that

$$c_1 \cdots c_{j-1}p = c_{j+1} \cdots c_{n-1}p'',$$

or $c_1 \cdots c_{j-1} = c_{j+1} \cdots c_{n-1}$ and $p = p''$, because C is prefix. Moreover, since $h(a_i) = c_i$ and h is injective on its alphabet, we have

$$a_1 \cdots a_{j-1} = a_{j+1} \cdots a_{n-1}.$$

We use this equality to show that w contains a square. Towards this end, notice that $h(a_0a_ja_n) = c_0c_jc_n = p'spsps''$ contains a square. It follows from the 3-square-free property of h that $a_0a_ja_n$ contains a square, thus $a_0 = a_j$ or $a_j = a_n$. In both cases, the word

$$w = a_0 \cdots a_n = a_0(a_1 \cdots a_{j-1})a_j(a_1 \cdots a_{j-1})a_n$$

contains a square, contradicting our assumptions on w. □

4. A 2-square-free test for square-freeness

Our next theorem is due to Pavel Goralčík and Tomas Vanicek. Before stating it, we need to introduce two more properties of codes in the spirit of infix.

Definition 3.14. A code $C \subseteq B^*$ is a **prefix-suffix code**, or a **ps-code**, if for every $c \in C$ and every factorization $c = ps$ in B^*, either c is the only word in C starting in p or c is the only word in C ending in s. That is,

$$ps, p's, ps' \in C \implies p' = p \text{ or } s' = s\,.$$

Note that we have allowed for trivial factorizations $ps = p\epsilon$ or $ps = \epsilon s$ above. In particular, ps-codes are bifix. The term ps-code was introduced in the thesis of Veikko Keränen. See also [Ker1986, Ker1987]. The notion has also been used by Michel Leconte (e.g., [Lec1985]), who calls such a code **faithful**.

Example. The codes $\{012, 120, 201\}$ and $\{012, 02122, 021102\}$ are ps-codes, while $\{0, 12, 102\}$ and $\{112, 0120, 012012\}$ are not.

Given a code $C \subseteq B^*$ and an element $w \in B^*$ (not necessarily in C), we say w is **left synchronizing** (in C) if for every $u, v \in B^*$ with $uwv \in C^*$, one has $u \in C^*$. The **right synchronizing** property is analogously defined. The property of codes we need is that of *strongly synchronizing*: a code C is called **strongly synchronizing** if for every $c \in C$ and every factorization $c = ps \in B^*$, p is left synchronizing or s is right synchronizing.

Remark. This may be compared to the more prevalent notion of **synchronizing** codes C: for all $c \in C$ and $ucv \in C^*$, one has $uc, cv \in C^*$. See [BP1985, Chapter VII.2] for more details. A strongly synchronizing code is comma-free and so synchronizing (Exercise 3.11). It is also infix (Lemma 3.13).

The condition of being strongly synchonizing seems rather difficult to check. Goralčík and Vanicek [GV1991] call a code **bissective** if it is a strongly synchronizing ps-code.

Theorem 3.15. *If a strongly synchronizing ps-morphism is* 2*-square-free, then it is square-free.*

Proof. A strongly synchronizing morphism $h : A^* \to B^*$ is infix (Exercise 3.11 and Lemma 3.13). If we can show that h is also 3-square-free, then Theorem 3.11 finishes the proof for us. Let C denote the code $C(h)$ and suppose we are given a word $w = a_1a_2a_3$ ($a_i \in A$) such that $h(w)$ contains the square uu for some $u \in B^* \setminus \{\epsilon\}$. We argue that w contains a square. Write $x = h(a_1)$, $y = h(a_2)$ and $z = h(a_3)$. We begin by reducing the problem to the situation illustrated in Figure 7.

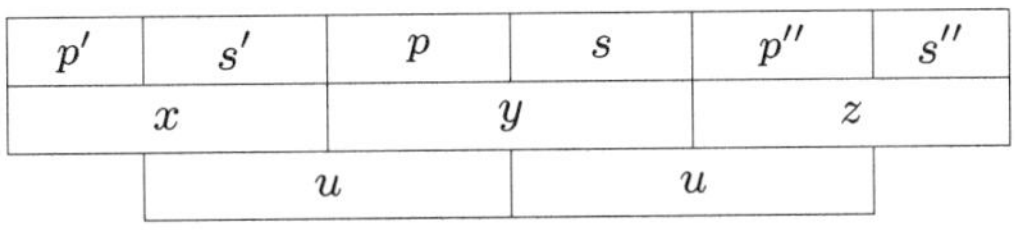

Figure 7. Factorizations of the code words $x, y, z \in C$ and the square $uu \in B^*$ inside the product xyz.

If uu is a factor of xy, then $h(a_1a_2) = xy$ contains a square. Since h is 2-square-free, a_1 must equal a_2 and w contains a square. Similarly, if uu is a factor of yz, then w contains a square.

Suppose uu is neither a factor of xy nor of yz. If u is a factor of x, then y is a factor of the second u, which implies y is a factor of x. Since C is infix, $x = y$, and w contains a square because h is 2-square-free. Similarly, if u is a factor of z, then w contains a square.

We thus have the situation depicted in Figure 7, with $s', p, s, p'' \neq \epsilon$. In particular, we learn that

$$xy = p's'ps = p'sp''s, \tag{3.16}$$

$$yz = psp''s'' = ps'ps''. \tag{3.17}$$

As h is strongly synchronizing and $y = ps \in C$, there are two cases to consider: p is left synchronizing or s is right synchronizing.

In the first case, (3.17) gives $ps' \in C^*$. So we have $ps \in C$ and $ps' = c_1 \cdots c_r \in C^*$ ($c_i \in C$). Since $psp'' = ps'p$, either c_1 is a prefix of ps or vice versa, both possibilities forcing $s' = s$ by the prefix property of C. Then (3.17) shows that $p'' = p$. In the second case, (3.16) gives $p''s \in C^*$, from which one also learns that $p'' = p$ and $s = s'$.

We now have

$$y = ps, \; x = p's, \; z = ps'' \in C,$$

which in turn implies either $p' = p$ or $s'' = s$ (as C is a ps-code). In other words $x = y$ or $z = y$, which shows that $w = xyz$ contains a square. □

Exercise 3.11. Prove that a strongly synchronizing code $C \subseteq B^*$ is both synchronizing and comma-free. (*Hint*: There are two cases to consider: either C is the entire alphabet B or it is not.)

CHAPTER 4

Squares in Words

This chapter deals with occurrences of squares in finite words. We begin by providing bounds for the number of occurrences of squares of primitive words and for the number of distinct squares in a fixed finite word. The remainder of the chapter is devoted to describing a linear-time algorithm to test whether a word contains a square. The necessary ingredients include centered squares, prefix arrays, the Crochemore factorization and suffix trees.

1. Counting squares

Here we establish bounds on the number of occurrences of squares of primitive words and the number of distinct squares in a fixed finite word.

Example. The word $aaaaaa$ contains three squares: aa, $aaaa$, $aaaaaa$. Only aa is a square of a primitive word, and it appears 5 times in $aaaaaa$.

Example. Let $w = abaababaabaab$. There are eight words whose squares occur in w: a; ab; ba; aba; baa; aab; $abaab$; and $baaba$. They are all primitive. The squares for a and aba occur thrice and twice, respectively.

Prefixes play an important role in dealing with squares, so we introduce the prefix poset of an alphabet.

Definition 4.1. Given an alphabet A, the **prefix poset** $\mathcal{P}_A = (A^*, <)$ is the poset defined by the order relation $x \leq y$ if x is a prefix of y. The poset contains a unique minimal element ϵ and is ranked by word length.

The explicit poset structure of $\mathcal{P}_A$ will play a role in Chapter 5. Here, we use only the ordering relation to help establish the desired bounds on squares in a word. The following result, due to Maxime Crochemore and Wojciech Rytter [CR1995, Lemma 10], will also be useful.

Lemma 4.2 (Three Squares Lemma). *Let u, v and w be words such that $uu < vv < ww$. If u is primitive, then $|u| + |v| \leq |w|$.*

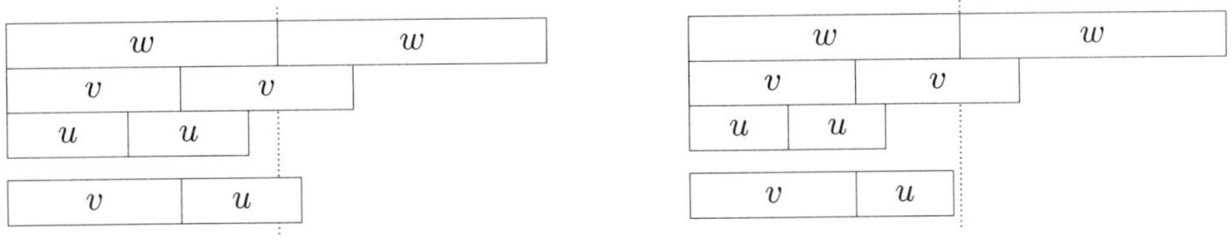

(a) w is a prefix of vu.

(b) vu is a prefix of w.

FIGURE 1. The Three Squares Lemma says (a) is impossible when u is primitive.

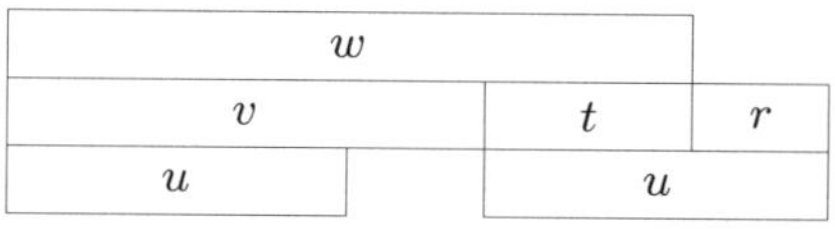

FIGURE 2. Proof of the Three Squares Lemma.

PROOF. Suppose u, v and w are three words such that $uu < vv < ww$ with $|w| < |u| + |v|$. We will show that u is not primitive. We begin by arguing for the situation illustrated in Figure 2.

Since $vv < ww$, we must have $v < w$ as well, so write $w = vt$ ($t \neq \epsilon$). Note that vt is a prefix of vv because both are prefixes of ww and $|vt| = |w| < |vv|$. It follows that t is a prefix of v. Since u is also a prefix of v, we have either $u \leq t$ or $t < u$. The first case would imply $|w| = |vt| = |t| + |v| \geq |u| + |v|$, contradicting our hypothesis. Therefore, $u = tr$ for some $r \neq \epsilon$. Finally, r is also a prefix of w because wr is a prefix of ww (since $wr = vtr = vu$ and vu is a prefix of vv, which is a prefix of ww).

Case I: $|u| + |t| > |v|$. Then vv is a prefix of wu, since $|wu| = |vtu| > |vv|$. Write $v = us$ for some nonempty word s. We are in the situation illustrated in Figure 3. We show that u is not primitive in 6 steps.

(1) *u begins with rs and sr. In particular, $sr = rs$.*

Since $wrs = vtrs = vus = vv$ and $vv < ww$, it follows that wrs is a prefix of ww of length $|vv|$. Also, wu is a prefix of ww and $|wu| > |vv|$. Hence, wrs is a prefix of wu, which implies that u begins with rs. So write $u = rsu'$. Then uu is a prefix of vu (since $uu < vv$ and $|u| < |v|$) and $vu = usu = usrsu'$. Hence, u is a prefix of $srsu'$, so u also begins with sr.

(2) *$r = z^\lambda$ and $s = z^\nu$ for some word z and some $\lambda, \nu \in \mathbb{N}$.*

This follows from Step 1 and Exercise 4.3.

(3) *$zu = uz'$ for some conjugate z' of z.*

Since $v = us$, it follows that uu is a prefix of $vu = usu$. So u is a prefix of su. Let s' be the word such that $su = us'$. Exercise 4.4 and Step 2 imply $zu = uz'$.

(4) *$ru = ur'$ for some word r'.*

This follows from Step 2 and Step 3.

(5) *$rt = tr$.*

Observe that vu is a prefix of wu since both words are prefixes of ww (because $u < v < w$) and $|vu| < |vv| < |wu|$. Thus vu is a prefix of vtu since $wu = vtu$. It follows that u is a prefix of tu. Write $tu = ut'$ for some word t'. Combined with Step 4, we have $rtu = rut' = ur't' = trr't'$.

(6) *u is not primitive.*

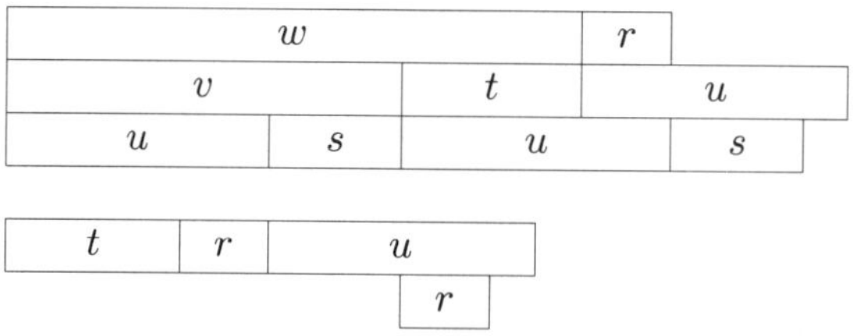

FIGURE 3. Case I in the proof of the Three Squares Lemma.

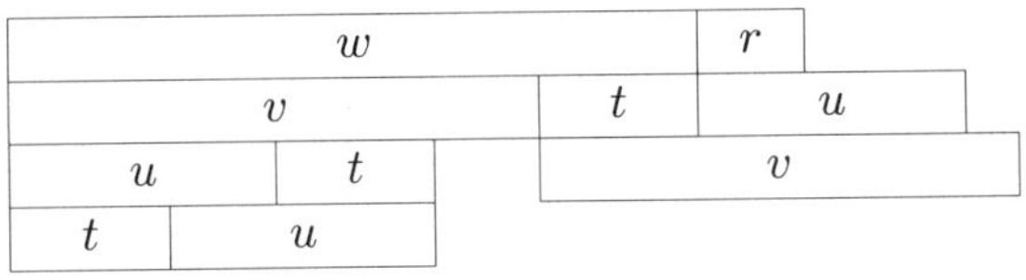

FIGURE 4. Case II in the proof of the Three Squares Lemma.

By Step 5 and Exercise 4.3, there exists a nonempty word p and nonnegative integers α and β such that $r = p^{\alpha}$ and $t = p^{\beta}$. Since r and t are nonempty, $\alpha + \beta \geq 2$. Hence, $u = tr = p^{\alpha+\beta}$ is not primitive.

Case II: $|u| + |t| \leq |v|$. This case is illustrated in Figure 4, and is argued as in Case I. The proof is left to Exercise 4.7. □

Remark. The primitive condition in the Three Squares Lemma is necessary: consider $u = a^3$, $v = a^4$ and $w = a^5$.

Our first application of the Three Squares Lemma concerns the number of squares of primitive words occurring in a finite word. This also appears in [CR1995].

For the convenience of the reader we recall the definition of big-O notation. Given two functions $f, g\colon \mathbb{N} \to \mathbb{R}_{>0}$, write $f(n) = O\big(g(n)\big)$ if there exist a positive integer N and a positive constant c such that $f(n) \leq cg(n)$ for every $n \geq N$. That is, $g(n)$ is an *asymptotic upper bound* for $f(n)$.

Theorem 4.3. *The number of occurrences of squares of primitive words in a word of length n is $O(n \log n)$.*

PROOF. We argue that the number of squares of primitive words that start in position i is always at most $\log_{\phi}(n) + 1$, where $\phi = (1+\sqrt{5})/2$ is the golden ratio.

Let w be a word of length n and let $y_1 < y_2 < \cdots < y_k$ be the primitive words such that $y_1^2, y_2^2, \ldots, y_k^2$ begin in position i. Repeated application of the Three Squares Lemma gives that $|y_j| \geq |y_{j-1}| + |y_{j-2}|$ for all $3 \leq j \leq k$. Since $|y_1| \geq 1$ and $|y_2| \geq 2$, it follows that $n = |w| \geq |y_k| \geq F_{k+1}$, where F_{k+1} is the $(k+1)$st Fibonacci number. This, in turn, is greater ϕ^{k-1} (see Exercise 4.9), thus $k \leq \log_{\phi}(n) + 1$. □

Remark. As shown in [Cro1981, Lemma 10], the bound provided in Theorem 4.3 is optimal and obtained by the prefixes of the Fibonacci word $\boldsymbol{f}$ of length F_k (the so-called "Fibonacci words," see Exercises 4.12 and 4.13).

Our second application of the Three Squares Lemma concerns the number of distinct squares in a word. It comes in the proof of the following theorem due to Aviezri S. Fraenkel and Jamie Simpson [FS1998].

Theorem 4.4. *Any word of length n contains at most $2n$ distinct squares.*

PROOF. Let m be a word of length n. For $0 \leq i \leq n-2$, denote by s_i the number of squares in m starting at position i that have no occurrence starting at a position greater than i. We are interested in the number $s_0 + s_1 + \cdots + s_{n-2}$. We will prove that $s_i \leq 2$ for each i.

Suppose on the contrary that $s_i \geq 3$. Then there exist three distinct words u, v and w with $uu < vv < ww$ such that uu, vv and ww begin at position i and have no occurrence starting at a position greater than i.

If u is primitive, then the Three Squares Lemma implies $|w| \geq |u|+|v| > 2|u| = |uu|$. Hence uu is a proper prefix of w. This implies that there is an occurrence of uu beginning at position $i + |w|$, a contradiction.

If u is not primitive, then $u = y^k$ for some primitive word y and some $k \geq 2$. So $yy < vv < ww$ since yy is a proper prefix of uu. By the Three Squares Lemma, $|w| \geq |v| + |y|$. We may assume that $|uu| > |w|$ because otherwise there is another occurrence of uu in m at a position greater than i. Therefore, w is a common prefix of y^{2k} and vv such that $|w| \geq |v| + |y| > |v| + |y| - \gcd(|v|, |y|)$. By the Fine–Wilf Theorem (Exercise 4.5), the words y and v are integer powers of some nonempty word. Since y is primitive, this word is y. Thus $v = y^\ell$ for some $\ell > k$. Therefore, $vv = y^{2\ell} = y^{2\ell-2k}y^{2k} = y^{2\ell-2k}uu$. Since vv begins at position i, it follows that uu also begins at position $i + (2\ell - 2k)|y| > i$. This is a contradiction. □

Remarks.

(1) Lucian Ilie recently proved a *three overlap lemma* similar in spirit to the Three Squares Lemma and used it to show that the number of squares occurring in a word of length n is bounded by $2n - O(\log n)$ [Ili2007].

(2) It has been conjectured that the $2n$ in the statement of Theorem 4.4 may be replaced by n [FS1998, Lot2005] [Ili2007]. In some sense, this is the best one can hope for: Fraenkel and Simpson construct a sequence of words—described in Exercise 4.8—where the number of squares in each word is very close to the length of the word.

Exercise 4.1 (Levi's Lemma). Let $u, v, x, y \in A^*$ and suppose $uv = xy$.

(a) If $|u| \geq |x|$, then there exists $t \in A^*$ such that $u = xt$ and $y = tv$.

(b) If $|u| < |x|$, then there exists $t \in A^* \setminus \{\epsilon\}$ such that $x = ut$ and $v = ty$.

Exercise 4.2. Let $y \in A^*$ and $x, z \in A^* \setminus \{\epsilon\}$. If $xy = yz$, then there exist words $u, v \in A^*$ and an integer $p \geq 0$ such that $x = uv$, $z = vu$ and $y = (uv)^p u = u(vu)^p$. (*Hint*: Proceed by induction on $|y|$ using Levi's Lemma above.)

Exercise 4.3. If $xy = yx$ for some words $x, y \in A^*$, then there exists a word $z \in A^*$ and nonnegative integers k and l such that $x = z^k$ and $y = z^l$. (*Hint*: Proceed by induction on $|xy|$.)

Exercise 4.4. Let x, y, z be words over some alphabet A. If $z^l x = xy$ for some positive integer l, then $zx = xz'$ for some conjugate z' of z.

Exercise 4.5 (Fine–Wilf Theorem). Let $u, v \in A^*$. There exists $w \in A^* \setminus \{\epsilon\}$ such that u, v are integer powers of w if and only if there exist $i, j \geq 0$ so that u^i and v^j have a common prefix (or suffix) of length $|u| + |v| - \gcd(|u|, |v|)$.

Exercise 4.6 (Synchronization). If u is a primitive word, then there are exactly two occurrences of u in the word uu (as a prefix and as a suffix).

Exercise 4.7. Complete Case II in the proof of Three Squares Lemma by arguing $|u| + |t| \leq |v|$ implies u is not primitive. (*Hint*: The situation is depicted in Figure 4; argue as in Case I.)

Exercise 4.8 ([FS1998]). For each $m \in \mathbb{N}$, let Q_m denote the concatenation of 00101001, 00010010001, …, and $0^{m+1}10^m10^{m+1}1$. For example, $Q_1 = 00101001$ and $Q_2 = 0010100100010010001$. Show that the number of squares having at least one occurrence in Q_m is very close to $|Q_m|$ by proving the following.

(a) The length of Q_m is $(3m^2 + 13m)/2$.
(b) The number of squares in Q_m is $(3m^2 + 7m - 6)/2 + \lfloor (m+1)/2 \rfloor$.

Exercise 4.9. Recall that the Fibonacci numbers are defined by $F_0 = 1$, $F_1 = 1$ and $F_n = F_{n-1} + F_{n-2}$ for all $n \geq 2$. Prove that $F_n > \phi^{n-2}$ for all $n \geq 2$, where $\phi = (1 + \sqrt{5})/2$. (*Hint*: Recall that ϕ satisfies a quadratic polynomial.)

2. Centered squares

The results in this and the next two sections may be found in the book by Maxime Crochemore, Christophe Hancart and Thierry Lecroq [CHL2001] and the English language reprint [CHL2007].

Given a factorization $w = (u, v) \in A^*$, a **centered square** at (u, v) is a factor $rsrs$ of w, with $rs \neq \epsilon$, such that either: $u = \alpha rsr$ and $v = s\beta$ for some $\alpha, \beta \in A^*$; or $u = \alpha r$ and $v = srs\beta$ for some $\alpha, \beta \in A^*$. See Figure 5. In the former case we say that $rsrs$ is a **left-centered square** of w at (u, v) and in the latter case we say that $rsrs$ is a **right-centered square** of w at (u, v).

In developing an algorithm for testing square-freeness, one could start with a "divide and conquer" method: given a word w, choose a factorization $w = (u, v)$ and look for centered squares at (u, v), as illustrated in Figure 5; if no centered square is found, then repeat the method for the words u and v. We begin by observing that no cavalier implementation of this method can possibly run in linear time.

Let $T_C(|u|, |v|)$ denote the time it takes to test for a left-centered square in uv at (u, v). Using the divide and conquer method, the computation time $T(n)$ for testing square-freeness of a word of length n is

$$T(n) = T\left(\left\lfloor \frac{n}{2} \right\rfloor\right) + T\left(\left\lceil \frac{n}{2} \right\rceil\right) + T_C\left(\left\lfloor \frac{n}{2} \right\rfloor, \left\lceil \frac{n}{2} \right\rceil\right).$$

We show in Section 3 that $T_C(|u|, |v|)$ is linear in $|uv|$, meaning that the divide and conquer algorithm only yields $T(n) = O(n \log n)$. Nevertheless, the notion of centered squares can be exploited to build a linear-time test for square-freeness. We establish this in Section 4.

Lemma 4.5. *A word uv has a left-centered square at (u, v) if and only if there exists a nontrivial factorization $u = (x, y)$ and words r and s such that*

(i) *r is a common suffix of u and x,*
(ii) *s is a common prefix of y and v,*
(iii) $|sr| \geq |y|$.

Exercise 4.10 asks the reader to turn Figure 6 into a proof of this lemma. The lemma becomes powerful with the observation that one is not looking for the beginning of the square.

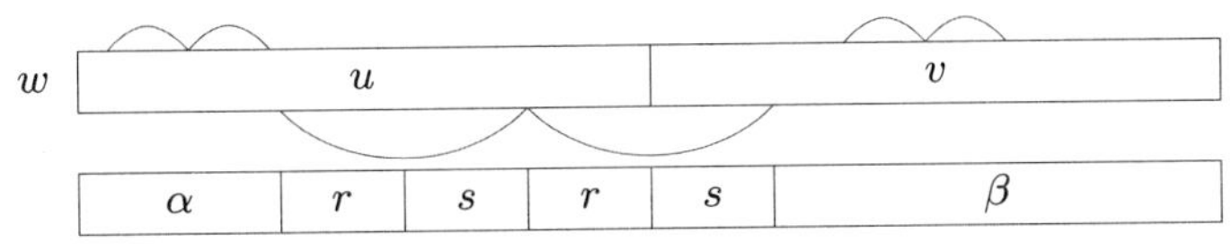

FIGURE 5. The divide and conquer method for finding squares.

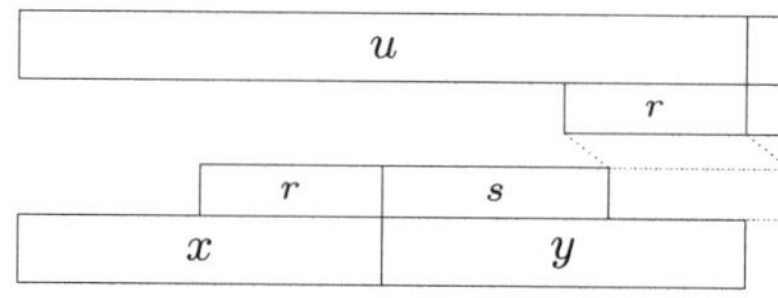

FIGURE 6. A picture proof of Lemma 4.5.

For a word $w = w_0 w_1 \cdots w_{n-1}$ of length n, we write $w(i,j)$ for the factor $w_i \cdots w_{j-1}$. Stopping short of w_j allows that $w(i,j)w(j,k) = w(i,k)$ and that $|w(i,j)| = j - i$. We abbreviate the prefix $w(0,i)$ of length i by $w_{(i)}$ and the corresponding suffix (starting at position i) by $w^{(i)}$ so that $w = w_{(i)} w^{(i)}$. Finally, we let $x \mathbin{/\!\!\!\wedge} y$ and $x \mathbin{\wedge\!\!\!\backslash} y$ denote, respectively, the **longest common prefix** and **longest common suffix** of x and y.

Corollary 4.6. *A word uv has a left-centered square at (u,v) if and only if there is an integer i $(0 \le i \le |u|-1)$ such that*

$$|u \mathbin{\wedge\!\!\!\backslash} u_{(i)}| + |v \mathbin{/\!\!\!\wedge} u^{(i)}| \ge |u| - i.$$

Exercise 4.10. Turn Figure 6 into a proof of Lemma 4.5.

3. Prefix arrays

The **prefix array** of two words x and y is the sequence of lengths of the longest prefixes common to x and suffixes of y:

$$\mathrm{pref}_{x,y}(i) := |x \mathbin{/\!\!\!\wedge} y^{(i)}|, \quad \text{for } 0 \le i \le |y|.$$

Similarly, the **suffix array** of x and y is the sequence of lengths of the longest suffixes common to x and prefixes of y:

$$\mathrm{suff}_{x,y}(i) := |x \mathbin{\wedge\!\!\!\backslash} y_{(i)}|, \quad \text{for } 0 \le i \le |y|.$$

Example. Consider the words $u = abacabaabacaa$ and $v = bacaccab$. We record a few prefix and suffix arrays in Figure 7.

Rephrasing Corollary 4.6 in this language, we see that a word uv has a left-centered square at (u,v) if and only if there exists $0 \le i \le |u|-1$ such that

$$\mathrm{suff}_{u,u}(i) + \mathrm{pref}_{v,u}(i) \ge |u| - i.$$

It is this formulation of the existence of left-centered squares that gives us $T_C(|u|,|v|) = O(|u|+|v|)$. Indeed, the complexity of computing $\mathrm{suff}_{u,u}$ and $\mathrm{pref}_{v,u}$

	0	1	2	3	4	5	6	7	8	9	10	11	12	13
u	a	b	a	c	a	b	a	a	b	a	c	a	a	
$\mathrm{pref}_{u,u}$	13	0	1	0	3	0	1	5	0	1	0	1	1	0
$\mathrm{suff}_{u,u}$	0	1	0	1	0	1	0	1	2	0	1	0	1	13
v	b	a	c	a	c	c	a	b						
$\mathrm{pref}_{v,u}$	0	4	0	0	0	2	0	0	4	0	0	0	0	0

FIGURE 7. Assorted prefix and suffix arrays for the words $u = abacabaabacaa$ and $v = bacaccab$.

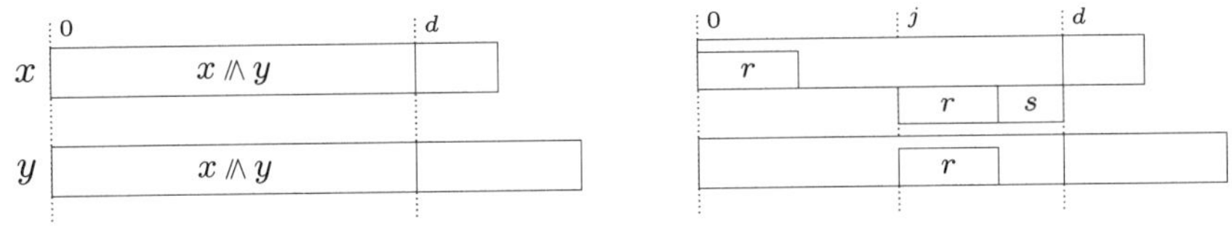

FIGURE 8. A proof of the first case of Lemma 4.7.

is linear in $|u|$ (we prove only the first of these facts here); likewise for right-centered squares and $|v|$.

Lemma 4.7. *Fix* $d = |x \wedge\!\!\wedge y|$. *For each* $0 < j < d$ *one has*

$$x \wedge\!\!\wedge y^{(j)} = \begin{cases} x \wedge\!\!\wedge x^{(j)} = y \wedge\!\!\wedge x^{(j)} = y \wedge\!\!\wedge y^{(j)}, & \text{if } |x \wedge\!\!\wedge x^{(j)}| < d - j, \\ x(j,d)\big(x^{(d-j)} \wedge\!\!\wedge y^{(d)}\big), & \text{if } |x \wedge\!\!\wedge x^{(j)}| \geq d - j. \end{cases}$$

PROOF. The case $|x \wedge\!\!\wedge x^{(j)}| < d - j$ is pictured in Figure 8 (the common prefix being r). From the picture, it is clear that the string of given quantities are all equal whenever $s \neq \epsilon$.

The argument supporting the other case is similar. □

Corollary 4.8. *Fix* $k < |x|$ *and* $d = \text{pref}_{x,x}(k)$. *For each* $0 < j < d$ *one has*

$$\text{pref}_{x,x}(k+j) = \begin{cases} \text{pref}_{x,x}(j), & \text{if } \text{pref}_{x,x}(j) < d - j, \\ d - j + |x^{(d-j)} \wedge\!\!\wedge x^{(k+d)}|, & \text{otherwise.} \end{cases}$$

The significance of this result is that one need not perform as many pref( ) and suff( ) calculations as one might guess. In Figure 9, we underline the prefix computations we get for free from the corollary. We leave it to the reader to develop the analogous suffix corollary and underline the free suffix computations.

We codify our findings as Algorithm 1. Note that the inner while-loop will not define $\mathsf{pref}[k+j]$ for values $k + j > n = |x|$ because d cannot be larger than $n - k$. In particular, the algorithm always terminates with the values $(k,d) = (n,0)$. For the reader unfamiliar with such computations, we carefully analyze the cost for one pass through Steps 3 – 11.

Suppose the pair (k,d) at Step 3 becomes (k',d') at Step 11. Then $k' - k - 1$ entries are simply copied from earlier in the array, for a "cost" on the order of $k' - k$, after which a single computation is performed with cost on the order of $|x^{(d-j)} \wedge\!\!\wedge x^{(k+d)}| = d' - d + j = d' - d + k' - k$. In total, we get $2(k'-k) + (d'-d)$

					$k=4$, $d=3$			$k=7$, $d=5$						
	0	1	2	3	[4]	5	6	(7)	8	9	10	11	12	13
x	a	b	a	c	a	b	a	a	b	a	c	a	a	
$\text{pref}_{x,x}$	13	0	1	0	[3]	<u>0</u>	1	(5)	<u>0</u>	<u>1</u>	<u>0</u>	1	1	0
$\text{suff}_{x,x}$	0	1	0	1	0	1	0	1	2	0	1	0	1	13
$\text{pref}_{x,x}(j) + j$		1	3	3	7									

FIGURE 9. Using Corollary 4.8 to compute the prefix array $\text{pref}_{x,x}$ of a word x. The underlined entries come at no cost.

```
input  : a word x of length n.
output: an array pref[0, n] of length n + 1.
pref[0] := n;   pref[1] := pref_{x,x}(1)
k := 1;   d := pref[k]
while k < n do
    j := 1
    while pref[j] < d − j do
        pref[k + j] := pref[j]
        j := j + 1
    end
    if d − j < 0 then d = j
    pref[k + j] := (d − j) + |x^{(d−j)} ∧ x^{(k+d)}|
    k := (k + j);   d := pref[k]
end
```

ALGORITHM 1. Computing the prefix array $\mathrm{pref}_{x,x}$ of a word x.

as the cost for one pass. If in r steps we move from $k_1 = 1$ to $k_r = n$, adding the cost of successive passes gives a telescoping sum. The total cost is on the order of $2k_r - 2k_1 - (d_r - d_1) = 2n - 2 - 0 + |x \wedge x^{(1)}|$, or $O(n)$.

Corollary 4.9. *For a given word x of length n, the prefix and suffix arrays $\mathrm{pref}_{x,x}$ and $\mathrm{suff}_{x,x}$ can be constructed in $O(n)$ time.*

Exercise 4.11 ([CHL2007, Theorem 2.35])**.** The complexity of computing $\mathrm{pref}_{v,u}$ is linear in $|u|$.

4. Crochemore factorization

Towards the goal of developing a linear-time test for square-freeness, Maxime Crochemore [Cro1983a] introduced a factorization of words similar to the popular Ziv–Lempel factorization.[1] His factorization of a word w, which may also be found in the literature as the "f-factorization" or "s-factorization," will be denoted $c(w)$ in what follows.

Definition 4.10. The **Crochemore factorization** of a word w is the unique factorization

$$c(w) = (x_1, x_2, \ldots, x_n)$$

of w with each x_i satisfying either:

(C1) x_i is a letter that does not appear in the factor $x_1 \cdots x_{i-1}$; or

(C2) x_i is the longest prefix of $x_i x_{i+1} \cdots x_n$ that also has an occurrence beginning within $x_1 \cdots x_{i-1}$ (i.e., there is a prefix ux_i of w with u shorter than $x_1 \cdots x_{i-1}$).

[1]The factorization introduced by Abraham Lempel and Jacob Ziv in [LZ1976] was later implemented by Terry Welch [Wel1984]. The so-called LZW algorithm is behind many lossless data compression algorithms (e.g., the TIFF image file format).

Example. The Crochemore factorizations of *abababb* and *abaababacabba* are

$$(\underset{1}{a}, \underset{1}{b}, \underset{2}{abab}, \underset{2}{b}) \quad \text{and} \quad (\underset{1}{a}, \underset{1}{b}, \underset{2}{a}, \underset{2}{aba}, \underset{2}{ba}, \underset{1}{c}, \underset{2}{ab}, \underset{2}{ba}),$$

where beneath each x_i in the factorization $(x_1, \ldots, x_n)$ we have written "1" or "2" according to whether (C1) or (C2) was used to build the factor.

The following result characterizes words containing squares in terms of its Crochemore factorization.

Notation. For any factor u of a word w, let $\pi_w(u)$ be the starting index of the first occurrence of u in w.

Theorem 4.11 (Crochemore [Cro1983a]). *Let w be a word with Crochemore factorization $c(w) = (x_1, \ldots, x_k)$. Then w contains a square if and only if there exists $j \in \mathbb{N}$ with $2 \leq j \leq k$ such that*

(i) $\pi_w(x_j) < |x_1 x_2 \cdots x_{j-1}| \leq \pi_w(x_j) + |x_j|$, *or*
(ii) *the pair* $(x_{j-1}\,, x_j)$ *has a centered square, or*
(iii) $j \geq 3$ *and the pair* $(x_1 \cdots x_{j-2}, x_{j-1}x_j)$ *has a right-centered square.*

Proof. Let $c(w) = (x_1, \ldots, x_k)$ be the Crochemore factorization of w. If (ii) or (iii) holds for some $2 \leq j \leq k$, then w obviously contains a square. Suppose (i) holds. This case is illustrated in Figure 10. Since $\pi_w(x_j) < |x_1 x_2 \cdots x_{j-1}|$, the first occurrence of x_j must begin within $x_1 x_2 \cdots x_{j-1}$. Option (A) in the figure is ruled out because it violates the condition $|x_1 x_2 \cdots x_{j-1}| \leq \pi_w(x_j) + |x_j|$. Options (B) and (C) provide, respectively, the squares $x_j x_j$ and rr within w.

Conversely, suppose that none of (i) – (iii) hold, yet there is a square zz within w. Letting j be minimal with zz a factor of $x_1 x_2 \cdots x_j$, we derive a contradiction in three steps.

(1) *zz is not a factor of x_j.*

Indeed, since (i) does not hold, either x_j is a new letter, in which case zz is clearly not a factor of x_j, or $\pi_w(x_j) + |x_j| < |x_1 x_2 \cdots x_{j-1}|$, meaning the first occurrence of x_j is a proper factor of $x_1 x_2 \cdots x_{j-1}$. Then zz is also a proper factor of $x_1 x_2 \cdots x_{j-1}$, contradicting the minimality of j.

(2) *zz is not a factor of $x_{j-1}x_j$.*

By Step 1, zz is not a factor of x_j. If zz is a factor of x_{j-1}, then we violate the minimality of j. Finally, if zz straddles the boundary between x_{j-1} and x_j, then (x_{j-1}, x_j) has a centered square, violating the assumption that (ii) does not hold.

(3) *zz is not a factor of $x_1 x_2 \cdots x_j$.*

After Steps 1 and 2, we need only rule out the case that zz is a centered square for $(x_1 x_2 \cdots x_{j-2}, x_{j-1}x_j)$. Since (iii) does not hold, we may assume zz is a left-centered square. We are left with the situation pictured in Figure 11 with $t \neq \epsilon$.

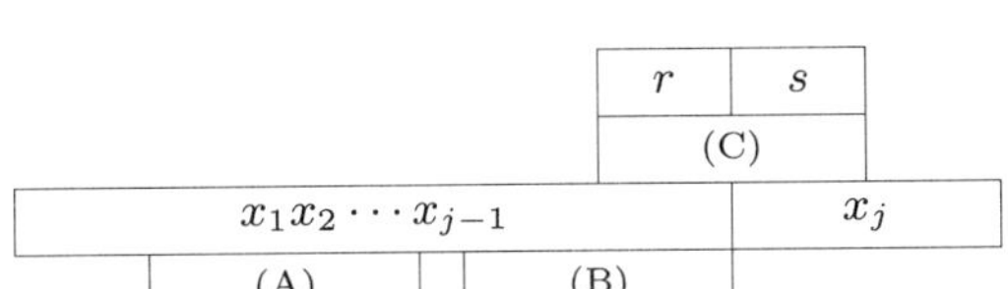

Figure 10. The possible positions (A)–(C) for the first occurrence of the word x_j within $x_1 x_2 \cdots x_{j-1} x_j$.

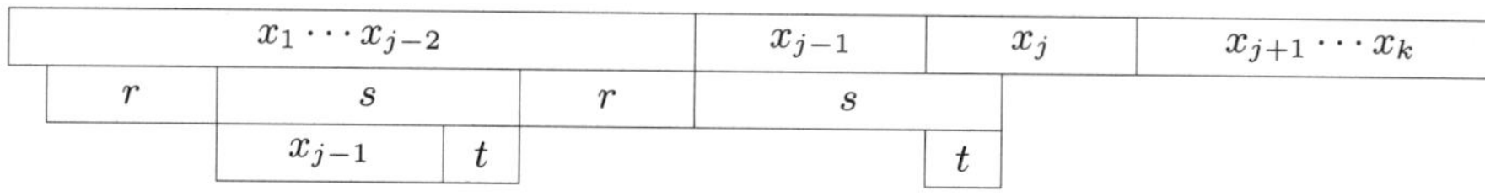

FIGURE 11. A left-centered square at $(x_1 \cdots x_{j-2}, x_{j-1}x_j)$ that satisfies none of conditions (i) – (iii) from Theorem 4.11.

By the definition of the Crochemore factorization, x_{j-1} cannot be a single letter (since it occurs before its indicated position as a prefix of s). On the other hand, (C2) is also violated: x_{j-1} should be the longest prefix of $x_{j-1} \cdots x_k$ that also has an occurrence within $x_1 \cdots x_{j-2}$, but s is strictly longer than x_{j-1}. □

We have already seen that the prefix (suffix) array for testing for left-centered (right-centered) squares can be built in linear time. Now, if the Crochemore factorization can be formed in linear time (and it can, see Section 5), then we have developed a square-free test that runs in linear time relative to $|w|$.

Corollary 4.12 ([CHL2007, Theorem 9.14]). *A word may be tested for square-freeness in linear time.*

PROOF. Given a word w to test, first compute the Crochemore factorization of w (which can be done in linear time; see Section 5), and then test Cases (i) – (iii) in Theorem 4.11 for $j = 2, \ldots, k$:

(i) *Is* $\pi_w(x_j) < |x_1 x_2 \cdots x_{j-1}| \leq \pi_w(x_j) + |x_j|$?
Each test takes $O(1)$ time.

(ii) *Does the pair* (x_{j-1}, x_j) *have a centered square?*
Each such test takes $O(|x_{j-1}|+|x_j|) = O(|x_{j-1}x_j|)$ time, by Corollaries 4.6 and 4.9.

(iii) *Does the pair* $(x_1 \cdots x_{j-2}, x_{j-1}x_j)$ *have a right-centered square?*
Each such test takes $O(|x_{j-1}x_j|)$ time because the cost of testing for a right-centered square at (u, v) is linear in $|v|$.

We conclude that test j may be completed in $O(|x_{j-1}x_j|)$ time. Summing this bound for each $j = 2, \ldots, k$, we find the total running time to be on the order of $O(|w|)$. □

Exercise 4.12 ([Smi1876]). Define an infinite word $\boldsymbol{f}$ as the limit of the iterations $f_0 = y$, $f_1 = x$, and $f_n = f_{n-1}f_{n-2}$ for $n \geq 2$. Verify that this is the Fibonacci word from Exercise 3.7. (The intermediate iterations in the procedure are often called the "Fibonacci words.")

Exercise 4.13 ([BS2006]). Compute the first six terms in the Crochemore factorization $(x_1, x_2, x_3, \ldots)$ of $\boldsymbol{f}$. Posit and prove a peculiar relationship between the x_is $(i \geq 4)$ and the finite Fibonacci words from Exercise 4.12.

5. Suffix trees

Given a word $w \in A^*$, the suffix tree $\mathcal{T}(w)$ is a data structure that compactly stores for fast retrieval every factor (in particular, every suffix) of w. Chief among its applications is a solution to the *factor problem.*[2]

[2]The terminology varies in the literature; this is also known as the *substring problem.*

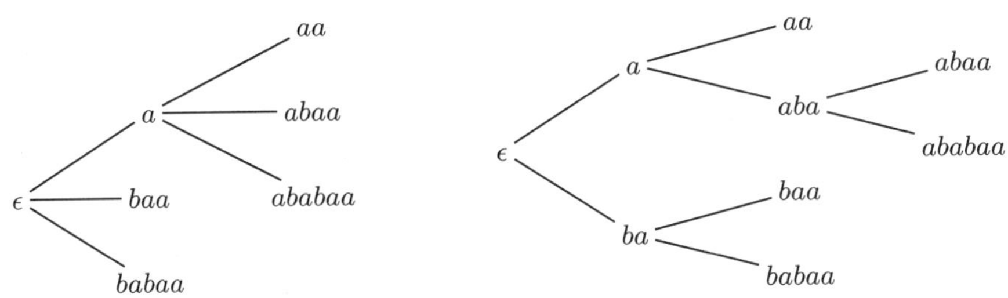

(a) The subposet Suff$(ababaa)$ in $\mathcal{P}_A$. (b) The closure $\overline{\text{Suff}}(ababaa)$ in $\mathcal{P}_A$.

FIGURE 12. Construction of the unadorned suffix tree $\overline{\text{Suff}}(w)$ for $w = ababaa$.

Factor Problem. *Given a word w of length n, find all locations of all occurrences of a factor f within w.*

It happens that $\mathcal{T}(w)$ can be built in $O(n)$ time (more precisely, $O(n \log |A|)$ time, but one typically assumes a fixed alphabet). Moreover, once this "preprocessing" is done, using $\mathcal{T}(w)$ one can solve the factor problem in $O(|f| + k)$ time, assuming k instances of the factor f within w. Otherwise stated, your favourite PDF viewer could find all instances of "Fibonacci" within the text you are reading faster than you could consult its index.

More information on the construction and applications of this "jewel" in stringology is readily found in [Gus1997, CR2002]. Here, we are chiefly concerned with its application to the Crochemore factorization. As such, we indicate the key ingredients of a linear time construction below but stop short of a detailed proof.

5.1. Definition and examples. The suffix tree may be described using the prefix poset $\mathcal{P}_A$ introduced in Definition 4.1. That poset has a *meet* operation[3] that we have already met: $(x, y) \mapsto x \wedge\!\!\wedge\, y$, the longest common prefix operation of Section 2. To begin, let Suff(w) denote the sub-poset of $\mathcal{P}_A$ consisting of the suffixes of w (including the empty word). The **unadorned suffix tree** of w, denoted by $\overline{\text{Suff}}(w)$, is the closure of Suff$(w)$ under $\wedge\!\!\wedge$ in $\mathcal{P}_A$.

Example. Figure 12 depicts the posets Suff$(ababaa)$ and $\overline{\text{Suff}}(ababaa)$.

Before defining the suffix tree, we need the notion of *covers:* if x and y are elements of a poset P, then we say that y **covers** x and write $x \lessdot y$, if $x < y$ and if there does not exist $z \in P$ such that $x < z < y$.

Definition 4.13. The **suffix tree** $\mathcal{T}(w)$ of a word w is the labelled, rooted, directed tree with nodes and arrows defined as follows.

Nodes: There is one node U for each $u \in \overline{\text{Suff}}(w)$. It carries two labels: the first occurrence $\pi_w(u)$ of u and "True" or "False" according to whether or not u is a suffix of w .

Arrows: There is a labelled arrow $U \xrightarrow{v'} V$ whenever U and V are the nodes corresponding to suffixes u and v with $u \lessdot v \in \overline{\text{Suff}}(w)$ and $v = uv'$.

The **root node** E is the node corresponding to ϵ. It is labelled 0 and True. A **suffix node** of $\mathcal{T}(w)$ is a node U labelled by True. A **leaf node** is a suffix node with no outgoing arrows.

[3]of Lattice theory, cf., [Sta1997, Chapter 3.3].

For every node U in $\mathcal{T}(w)$, let POSITION(U) be the numerical label attached to U, let SUFFIX(U) denote the Boolean label attached to U and let PATH$(U) = v_1v_2\cdots v_t$, where $E \xrightarrow{v_1} \cdots \xrightarrow{v_t} U$ is the unique path in $\mathcal{T}(w)$ from E to U. Then PATH(U) is the word $u \in \overline{\mathrm{Suff}}(w)$ corresponding to the node U, SUFFIX(U) is True if and only if PATH$(U) \in \mathrm{Suff}(w)$, and POSITION$(U) = \pi_w(\text{PATH}(U))$.

Notation. In the interest of legibility, we keep the labelling convention introduced above by representing nodes with capital letters and their associated paths with the corresponding lowercase letter, e.g., U is the node corresponding to $u \in \overline{\mathrm{Suff}}(w)$. In figures, the position property of a node is displayed, and the node is double-circled if it is a suffix node.

Example. The suffix tree $\mathcal{T}(ababaa)$ is pictured in Figure 13. (Compare it with Figure 12(b).)

Example. The suffix tree $\mathcal{T}(abacabaabaca)$ is pictured in Figure 14.

Remarks.

(1) In the preceding examples, each factor f appearing in w is a prefix of PATH(U) for a unique node U in $\mathcal{T}(w)$. This is true in general. See Exercise 4.14.

(2) The above construction is often called the **compact suffix tree** in the literature, in which case the name "suffix tree" is reserved for the tree $\mathcal{T}^\sharp(w)$ that

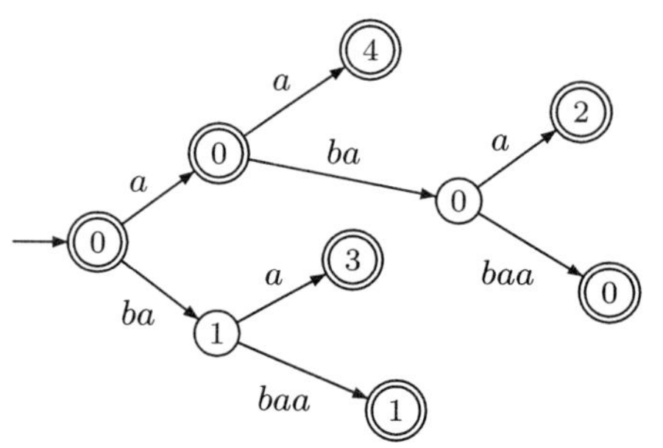

FIGURE 13. The suffix tree $\mathcal{T}(ababaa)$.

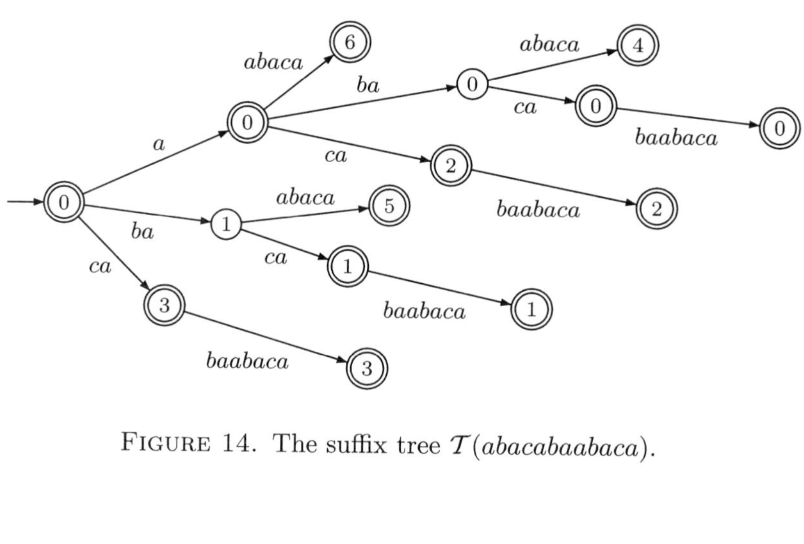

FIGURE 14. The suffix tree $\mathcal{T}(abacabaabaca)$.

may be recovered from $\mathcal{T}(w)$ by adding enough bivalent nodes so that all arrow labels are letters. The advantage of using $\mathcal{T}^\sharp(w)$ is that it may be viewed as an automaton whose language is $\mathrm{Suff}(w)$. The disadvantage is that it is too large (in terms of the number of states and arrows).

Proposition 4.14. *If a tree $\mathcal{T}$ has n nodes and e edges, let $n+e$ be a measure of the* size *of $\mathcal{T}$. Given a word $w \in A^*$ of length N, the trees $\mathcal{T}(w)$ and $\mathcal{T}^\sharp(w)$ have sizes $O(N)$ and $O(N^2)$, respectively.*

PROOF. Exercise 4.15. □

In particular, since the run-time for an algorithm is at least as large as its output, constructing $\mathcal{T}^\sharp(w)$ in linear time is impossible.[4]

The suffix tree for w may be built recursively in several different ways. Peter Weiner [Wei1973] and Edward M. McCreight [McC1976] were the first to describe algorithms to do this in linear time. Weiner's method adds longer and longer suffixes of w to a tree, starting with the last letter of w, whereas McCreight's method adds shorter and shorter suffixes, starting with the entire word w. For more details on their algorithms and how they might be implemented in linear time, see [Gus1997] and [CHL2007] respectively.

Below, we briefly describe the "on-line" algorithm of Esko Ukkonen [Ukk1995] for constructing $\mathcal{T}(w)$. Ukkonen's method may also be implemented in $O(|w| \log |A|)$ time. It has the novel advantage of not needing to see the entire word w before beginning the construction of $\mathcal{T}(w)$. Our description largely follows that of [Gus1997].

5.2. On-line construction. Recall the notation $w_{(i)}$ and $w(j,i)$ from Section 2 for the prefix $w_0 \cdots w_{i-1}$ and the factor $w_j \cdots w_{i-1}$, respectively, of $w = w_0 \cdots w_{n-1}$. The idea driving Ukkonen's on-line algorithm is to start by building the suffix tree $\mathcal{T}(w_{(1)})$, then continue recursively, building the tree $\mathcal{T}(w_{(i+1)})$ from $\mathcal{T}(w_{(i)})$ for $1 \leq i < n$. The passage from one phase of the construction to the next exploits a simple relationship between the suffixes of $w_{(i+1)}$ and $w_{(i)}$:

$$\begin{aligned}\mathrm{Suff}(w_{(i+1)}) &= \{uw_i \mid u \in \mathrm{Suff}(w_{(i)})\} \cup \{\epsilon\} \\ &= \{w(j,i)w_i \mid 0 \leq j \leq i\} \cup \{\epsilon\}.\end{aligned}$$

In order to achieve the $O(n)$ complexity, Ukkonen actually works with *implicit suffix trees* at each step, passing to a true suffix tree only at the last step. The **implicit suffix tree** $\mathcal{T}^\flat(w)$ of a word w differs from $\mathcal{T}(w)$ in that the root node and interior suffix nodes are not labelled as suffixes, and interior bivalent nodes do not appear at all.

Example. Figure 15 depicts the implicit version of the tree in Figure 14.

We begin by constructing $\mathcal{T}^\flat(w)$. As we will see shortly, one easily passes from $\mathcal{T}^\flat(w)$ to $\mathcal{T}(w)$ by one extra walk through the tree. Let us say that there are n *phases* in the construction, with phase $i+1$ corresponding to adding the letter w_i to the tree $\mathcal{T}^\flat(w_{(i)})$. The reader may find it useful to look at the elementary examples in Figure 16 before reading further. The first phase is always the same and handled in constant time: make the tree $E \xrightarrow{w_0} U$ with $\mathrm{POSITION}(E) = 0$, $\mathrm{SUFFIX}(E) =$ "False", $\mathrm{POSITION}(U) = 0$ and $\mathrm{SUFFIX}(U) =$ "True".

[4] There does exist a general scheme for building an automaton of smallest size with language $\mathrm{Suff}(w)$. We refer the interested reader to [CHL2007, Chapter 5.5] for details.

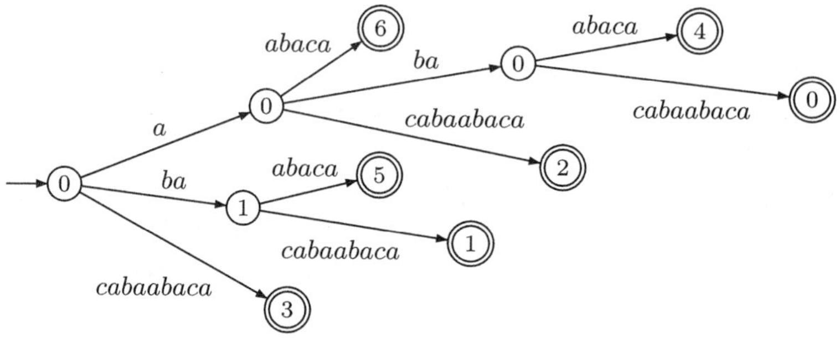

FIGURE 15. The implicit suffix tree $\mathcal{T}^\flat(abacabaabaca)$.

Within phase $i+1$ (for $i > 1$), there are $i+1$ *extensions,* one for each suffix of $w_{(i)}$ (incuding the empty suffix). We order the suffixes from longest to shortest, so extension $j+1$ corresponds to processing the factor $w(j,i)w_i$. Each extension is processed according to rules below.

Rule 1. If $v = uv'$ is a suffix of $w_{(i)}$ corresponding to a leaf node V, and V is a leaf node of U with $\text{PATH}(U) = u$, then replace the arrow $U \xrightarrow{v'} V$ with $U \xrightarrow{v'w_i} V$.

Rule 2. If $u = w(j,i)$ is a suffix of $w_{(i)}$ corresponding to a node U of $\mathcal{T}^\flat(w_{(i)})$ and no edge leaving U begins by w_i, then add a new node V: set $\text{POSITION}(V) = j$, $\text{SUFFIX}(V) =$ True and add the edge $U \xrightarrow{w_i} V$.

Rule 3. If $u = w(j,i)$ is a suffix of $w_{(i)}$ that does not correspond to a node of $\mathcal{T}^\flat(w_{(i)})$, i.e., there are nodes $V_1 \xrightarrow{v} V_2$ with $u = \text{PATH}(V_1)u'$ and $v = u'v'$, and if v' does not begin by w_i, then add two new nodes U and V: set $\text{SUFFIX}(U) =$ False, $\text{SUFFIX}(V) =$ True, $\text{POSITION}(V) = j$ and $\text{POSITION}(U) = \text{POSITION}(V_2)$; add the edges $U \xrightarrow{w_i} V$ and $U \xrightarrow{v'} V_2$ and replace $V_1 \xrightarrow{v} V_2$ by $V_1 \xrightarrow{u'} U$.

Rule 4. If $u = w(j,i)$ is a suffix of $w_{(i)}$ and a path extending from u begins by w_i, then *do nothing* (the suffix $w(j,i)w_i$ is already present).

The sequences $(a_0, a_1, \dots)$ in Figure 16 indicate which rules were applied during phase i. One might observe the following behaviour (Exercise 4.16).

Lemma 4.15. *Let $a_j(i)$ denote the rule used during extension j of phase i. Then:*

(i) *if $a_j(i) \in \{1,2,3\}$, then $a_j(i') = 1$ for all future phases i';*
(ii) *if $a_j(i) = 4$, then $a_{j'}(i) = 4$ for future extensions j' of i.*

The importance of the lemma will be evident a bit later. First, let us explain how to recover the suffix tree $\mathcal{T}(w)$ from $\mathcal{T}^\flat(w)$ by processing an additional 'end-of-text' character external to the alphabet A (say \$). Note that since \$ does not appear in A, no suffix of $w\$$ is a proper prefix of another (i.e., there are no implicit suffix nodes in $\mathcal{T}^\flat(w\$)$). Instead of using Rules 1–4 to pass from $\mathcal{T}^\flat(w)$ to $\mathcal{T}^\flat(w\$)$, we use the following modifications and arrive directly at $\mathcal{T}(w)$: if Rule 1 should apply for $w(j,n)\$$, do nothing; if Rule 2 or 3 should apply, set $\text{SUFFIX}(U) =$ "True" in place of adding the new node V; Rule 4 will never apply by our choice of \$.

Proposition 4.16. *The above procedure constructs the suffix tree $\mathcal{T}(w)$ of a word $w \in A^*$ in $O(|w|^3)$ time.*

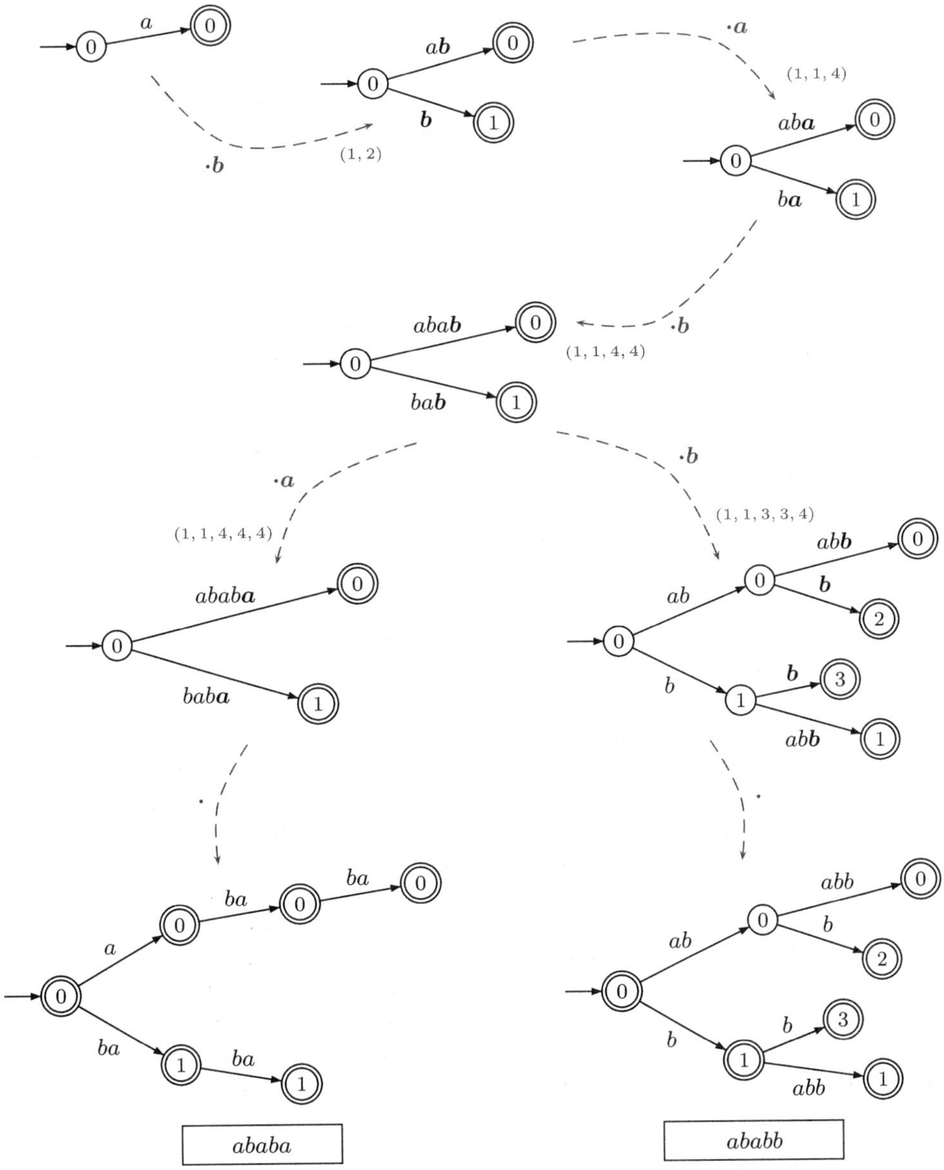

FIGURE 16. The suffix trees $\mathcal{T}(ababa)$ and $\mathcal{T}(ababb)$, built using Ukkonen's on-line algorithm. Each implicit suffix tree is built from the last by lengthening leaves or adding nodes in a systematic manner.

PROOF. The key issue is how to locate the ends of the $i+1$ suffixes of $w_{(i)}$ within $\mathcal{T}^\flat(w_{(i)})$. If we start from the root of the current tree, the jth extension to phase $i+1$ would take $O(i+1-j)$ time to locate. We leave the details to Exercise 4.17. □

5.3. Towards a linear-time algorithm. So far, we have a blueprint for construction of $\mathcal{T}(w)$ in $O(|w|^3)$ time—hardly the $O(|w|)$ time advertised in the introduction. Ukkonen begins with this blueprint and introduces several modifications to reduce the time and space demands of the execution. Chief among them is the notion of *suffix links* that stems from the following observation, which we leave as an exercise.

Lemma 4.17. *Suppose $w \in A^*$. If U is a node in $\mathcal{T}^\flat(w)$ with* PATH$(U) = av$ *for some $a \in A$, then there exists a unique node V with* PATH$(V) = v$.

Define the **suffix link** function S on nodes of $\mathcal{T}^\flat(\ )$ by $S(U) = V$ (in the notation of Lemma 4.17). Note that once $S(U)$ has been defined in phase i, its value does not change in subsequent phases. Less trivial is the observation that if a new node U is created in extension j of some phase i, then it will be possible to define $S(U)$ at least by the end of extension $j+1$ (so $S(U)$ is ready for use in phase $i+1$). Creating and maintaining suffix links greatly reduces the time required to apply Rules 1–4 during each phase. Indeed, if during phase i you have finished working on extension j, you need not march from the root all the way to $w(j+1,i)$. You can get there in constant time if $w(j,i)$ corresponds to a node (following a suffix link), and in time proportional to $i-j-1$ otherwise. Implemention of suffix links thus reduces the total run-time from $O(|w|^3)$ to $O(|w|^2)$.

We are now ready to use Lemma 4.15. The next speed boost begins by reducing the space requirements, replacing the edge labels $w(j,k)$ with the pair of integers (j,k) (see Exercise 4.15). This is useful as follows. If $w(j,i)$ corresponds to a leaf, it will correspond to a leaf for all $i' > i$ by the lemma. As a result, it is sufficient to just label it (j,e) where e represents the "current-end-of-text." Incrementing e once per phase is sufficient to deal with all leaves in that phase. Moreover, one may simply keep a counter j_0 that points to the "last-known-leaf" from the previous phase. Then j_0+1 is the starting index of the first suffix in the current phase that needs to be explicitly dealt with. Next, again by the lemma, one notes that as soon as Rule 4 is applied in extension j, one can immediately skip to the start of the next phase (all remaining extensions j' will be by Rule 4 and all relevant suffix links have already been made). Finally, Ukkonen shows that the total number of times Rule 2 or 3 must be applied is bounded by $2|w|$. Implementing these last tricks yields the advertised $O(|w|)$ time for the on-line algorithm. Again, we refer the reader to [Gus1997] for omitted details.

5.4. Crochemore factorizations and square-free tests. We have shown that using prefix arrays and the Crochemore factorization $c(w)$ of a word w, one can test for square-freeness in $O(|w|)$ time. To verify that the test can truly run in $O(|w|)$ time, it remains only to show that $c(w)$ can be built in linear time.

Proposition 4.18 (Crochemore [Cro1986]). *The Crochemore factorization $c(w) = (x_1, x_2, \dots, x_n)$ of a word w may be realized in linear time.*

PROOF. As usual, the factor x_1 is the single letter w_0. Next, assume that the first i factors $x_1, x_2, \dots, x_i$ have been found and write $w = x_1 \cdots x_i x'$ with

$|x_1 \cdots x_i| = N$. Use the suffix tree to search for prefixes p of x' within w. Choose p such that $|p|$ is maximal among all prefixes q with $\pi_w(q) < N$. If $p = \epsilon$, then put x_{i+1} equal to w_N (the first letter of x'). Otherwise, put $x_{i+1} = p$.

The above procedure evidently results in the Crochemore factorization of w. Since the suffix tree can be built in linear time, it remains only to verify that the indicated search may be carried out in linear time. We leave this to Exercise 4.20. □

Example. The Crochemore factorization $(a, b, a, c, aba, abaca)$ is readily built from the suffix tree $\mathcal{T}(abacabaabaca)$ in Figure 14. It passes Tests (i) and (iii) of Theorem 4.11 at each index $2 \leq j \leq 6$ and first fails test (ii) at (x_5, x_6).

Complete details on the use of Crochemore's algorithm to test for square-freeness may be found in [CHL2007, Chapter 9.3]. We mention that Dan Gusfield and Jens Stoye have a suffix-tree method for testing square-freeness that avoids both the Crochemore factorization and suffix arrays [SG2002].[5] While their algorithm is somewhat simpler to describe, the present one affords us the opportunity to introduce more tools of the trade.

5.5. Further applications. It is worthwhile to mention a few additional uses of suffix trees. Exercises 4.18 – 4.22 outline several that are within reach, despite the rapid introduction to suffix trees above. Touching on a central theme of Part I, suffix trees may also be used to find the maximal palindrome p within a word w in linear time. We close by mentioning the k-mismatch problem.

Fix two words $s = s_0 \cdots s_r$ and $w = w_0 \cdots w_n$ over an alphabet A, with $r \leq n$. Given $k \in \mathbb{N}$, a **k-mismatch** of s within w is a factor $w_i \cdots w_{i+r}$ of w such that $w_{i+j} \neq s_j$ for at most k integers $0 \leq j \leq r$. The k-mismatch problem is to find all k-mismatches of s within w. Gad M. Landau and Uzi Vishkin [LV1986] and Eugene W. Myers [Mye1986] have shown that this can be done in $O(kn)$ time. Moreover, Gad M. Landau and Jeanette P. Schmidt [LS1993] developed a method to test for k-mismatch square-freeness in $O\big(kn\log(n/k)\big)$ time. It uses, as we did in Section 2, a divide-and-conquer idea as its starting point.

Exercise 4.14. Verify from the definition of $\mathcal{T}(w)$ that if f is any proper factor of w, then there is a unique node $U \in \mathcal{T}(w)$ with f a prefix of $\textsc{path}(U)$.

Exercise 4.15. Prove Proposition 4.14. Also, prove that if one replaces the factors $w(j, k)$ labelling edges by the extremal indices (j, k), then $\mathcal{T}(w)$ also takes up only $O(w)$ space.

Exercise 4.16. Prove Lemma 4.15.

Exercise 4.17. Prove Proposition 4.16.

Exercise 4.18 (The Factor Problem). Given a text w and a factor f, return in $O(|w| + |f| + k)$ time all k instances of f within w. (*Hint*: Assume, or prove, that the following modification of $\mathcal{T}(w)$ may also be built in $O(|w|)$ time: instead of labelling internal nodes U with $\textsc{position}(U)$, label them with the set $\{\textsc{position}(V) \mid V \text{ is a leaf node whose path passes through } U\}$.)

[5] Though admittedly they introduce so-called *branching squares* and *DFS* arrays in their place.

Exercise 4.19 (Longest Common Factor). The longest common factor of two words u, v can be found in linear time relative to $|u| + |v|$. (*Hint*: Build a suffix tree for each word first, and do a bottom-up search. See [CR2002, Theorem 5.3].)

Exercise 4.20 (Longest Prefix Factor). Given the suffix tree for a word $w \in A^*$ and a suffix s beginning at position i in w, return in $O(|p|)$ time the longest prefix p of s such that $\pi_w(p) < i$.

Exercise 4.21 (All Distinct Factors). The number of distinct factors of a word w may be counted in $O(|w|)$ time. The distinct factors may be enumerated (printed) in $O(|w| + N)$ time, where N is the sum of the lengths of the distinct factors of w.

Exercise 4.22 (Smallest k-Repeat). Given $w \in A^*$ and $k \in \mathbb{N}$, find in $O(|w|)$ time the shortest factor s that occurs exactly k times within w.

CHAPTER 5

Repetitions and Patterns

In this final chapter, we outline several directions of research departing from the Thue–Morse word and square-free work of the preceding chapters. More details may be found in [Lot1997, Chapter 2], [Lot2002, Chapter 3] and, of course, the original sources cited below.

1. Maximal repetitions

Let v and w be words and write $w = w_0w_1w_2\cdots$, where $w_0, w_1, w_2, \ldots$ are letters. Recall that $w_iw_{i+1}\cdots w_j$ is said to be an **occurrence** of v in w if $v = w_iw_{i+1}\cdots w_j$. (In particular, an occurrence of v in w includes information about where v appears in w.) Here we will be concerned with occurrences of *repetitions* in a fixed word.

A **repetition** is a word of the form u^nv, where u is a word, v is a prefix of u and $n \geq 2$. Thus, squares and overlaps are examples of repetitions. If r is a repetition, then the **minimal period** of r is the smallest positive integer p such that $r_i = r_{i+p}$ for all $0 \leq i < |r|-p$. An occurrence of a repetition $r = w_i \cdots w_j$ in w is **extendible** if $w_{i-1}r$ or rw_{j+1} has the same minimal period as r. An occurrence of a repetition is **nonextendible** if it is not extendible. A **maximal repetition**, or a **run**, in a word is an occurrence of a nonextendible repetition in the word.

Examples. Let $w = abaabababaaab$.

(1) The second occurrence of *baba* in w is an extendible repetition since *ababa* has minimal period 2. Note that *babaa* does not have minimal period 2. The first occurrence of *baba* in w is also an extendible repetition.

(2) The first occurrence of the square *aa* in w is a maximal repetition, but the other two occurrences of *aa* are not maximal repetitions because they occur in the cube *aaa*.

(3) The occurrences of *abababa*, *aaa* and *abaaba* in w are also maximal repetitions.

Any occurrence of a repetition in a word w can be extended to a maximal repetition, so the maximal repetitions in w carry all the information about repetitions that occur in w. Moreover, thanks to the work of Roman Kolpakov and Gregory Kucherov [KK1999a, KK1999b], we know that the number of maximal repetitions in a word of length n is linear in n, so this fact has practical applications. However, their arguments did not provide an explicit bound on the number of maximal repetitions. The first such bound, $5n$, was established by Wojciech Rytter [Ryt2006], and was later improved to $3.44n$ [Ryt2007]. The best result in this direction was recently announced by Maxime Crochemore and Lucian Ilie.

Theorem 5.1 (Crochemore, Ilie [CI2007]). *The number of maximal repetitions in a word of length n is less than $1.6n$.*

Crochemore and Ilie also provide suggestions on how to improve the above bound [CI2007, Section 7]. In their analysis, they use the fact that the number of maximal repetitions with periods at most 9 in a word of length n is at most n (see their Lemma 2). If this fact can be improved, then their argument gives a better bound. For example, if it holds for periods at most 32, then their argument gives a bound of $1.18n$.

2. Repetition thresholds

In addition to minimal periods $p = |u|$ for the words $r = u^n v$ above, there is the notion of **exponent** $|r|/p$, a number lying between n and $n+1$ and measuring how close r is to a pure power. Given an infinite word $\boldsymbol{s}$, the **critical exponent** of $\boldsymbol{s}$ is the supremum of the exponents of all its (finite) factors r.

Françoise Dejean [Dej1972] introduced this notion in the interest of generalizing the square-free questions of Thue. She asked,

> *Given a fixed n-letter alphabet A, what is the minimal critical exponent among all infinite words over A?*

We call this the **repetition threshold** $\mathrm{RT}(n)$ for n.

Example. A quick check reveals that every binary word of length four has a factor with exponent 2. Hence, the exponent is at least 2 for any infinite binary word $\boldsymbol{s}$. On the other hand, Exercise 5.1 tells us that the exponent of the Thue–Morse word $\boldsymbol{t}$ is at most 2. That is, $\mathrm{RT}(2) = 2$.

Appealing to Thue's work again, we recall that there exist square-free words on 3 letters, so $\mathrm{RT}(3) < 2$. Dejean was able to show that, in fact, $\mathrm{RT}(3) = 7/4$. She was also able to show that $\mathrm{RT}(4) \geq 7/5$ and $\mathrm{RT}(n) \geq n/(n-1)$ for $n \geq 5$. She furthermore conjectured that these bounds are tight.

In [Pan1984], Jean-Jacques Pansiot finally answered the question in the affirmative for $n = 4$. Results for larger n have been less forthcoming. Jean Moulin-Ollagnier verified the case $5 \leq n \leq 11$ [Mou1992]. Interestingly, James D. Currie and Mortez Mohammad-Noori verified the cases $12 \leq n \leq 14$ [CM2007] by using certain Sturmian words to construct binary words whose critical exponents achieve the desired lower bound $n/(n-1)$. The work continues and is likely to end in a positive answer. For example, Arturo Carpi has proven that Dejean's conjecture holds for all $n \geq 33$ [Car2006, Car2007].

Exercise 5.1. Words of the form $r = u^n v$, with v a prefix of u, are called **fractional powers** in the literature.[1] Prove that a word w is overlap-free if and only if it contains no fractional power r as a factor with exponent greater than 2.

Exercise 5.2 ([BBGL2007])**.** Suppose $\beta \geq 2$ and $m \geq 1$ are integers. Let A be an m-letter alphabet A and fix $a \in A$ and a cyclic permutation σ of A. Define a generalization of the Thue–Morse word as $\boldsymbol{t}_{\beta,m} = \xi^\infty(a)$, where

$$\xi(x) = x\sigma(x)\sigma^2(x)\cdots\sigma^{\beta-1}(x)$$

[1]One can find these called "sesquipowers" in the literature, even for $n > 1$. We prefer to reserve that terminology for $n = 1$, given the etymology of "sesqui."

for all $x \in A$. Show that the critical exponent $e(\boldsymbol{t}_{\beta,m})$ of $\boldsymbol{t}_{\beta,m}$ is

$$e(\boldsymbol{t}_{\beta,m}) = \begin{cases} \infty, & \text{if } m \mid \beta - 1, \\ 2\beta/m, & \text{if } m \nmid (\beta - 1) \text{ and } \beta > m, \\ 2, & \text{if } \beta \leq m. \end{cases}$$

3. Patterns

A **pattern** p is a nonempty word over some alphabet A. Given a pattern $p = a_1 a_2 \cdots a_n$ ($a_i \in A$), an **instance of a pattern** p is a word $x_1 x_2 \cdots x_n$ ($x_i \in B^* \setminus \{\epsilon\}$) such that $a_i = a_j$ implies that $x_i = x_j$. Equivalently, $x_1 x_2 \cdots x_n$ is an instance of p if there exists a nonerasing morphism $h\colon A^* \to B^*$ such that $h(p) = x_1 x_2 \cdots x_n$. An **Abelian instance** of a pattern p is a word $x_1 x_2 \cdots x_n$ such that $a_i = a_j$ implies that x_i equals some rearrangement, or **anagram** of x_j, i.e., $|x_i|_b = |x_j|_b$ for each $b \in B$. We denote this by $x_i \sim x_j$.

Example. Consider the pattern $p = aabb$. The word

$$zxzyzxzyzxyyxyyyz = z(xzyz)(xzyz)(xyy)(xyy)yz$$

contains an instance of p (use the mapping $(a, b) \mapsto (xzyz, xyy)$), while

$$xyzxxyzxzyyyxyzy = x(yzx)(xyz)(xzyy)(yxyz)y$$

contains an Abelian instance of p.

Definition 5.2. A pattern p is **k-avoidable** (respectively, **Abelian k-avoidable**) if there exists an infinite word $\boldsymbol{x}$ on k letters that contains no instance (Abelian instance) of p as a factor.

Remarks.

(1) Note that a word on k letters is also a word on $k + 1$ letters, i.e., "k-avoidable" implies "$(k + 1)$-avoidable" for all $k \in \mathbb{N} \setminus \{0\}$.

(2) In terms of morphisms, we say that p is k-avoidable if and only if for all nonerasing morphisms $h\colon A^* \to B^*$, $h(p)$ is not a factor of $\boldsymbol{x}$. Likewise for Abelian k-avoidable.

Examples.

(1) The pattern $ababa$ is 2-avoidable since $\boldsymbol{t}$ is overlap-free.

(2) The pattern aa is 3-avoidable because there exists an infinite word on 3 letters that is square-free (e.g., the word $\boldsymbol{m} = 2102012\cdots$ from Section 1 of Chapter 3).

The study of words avoiding patterns goes back to Thue [Thu1906]. He asked, given $p \in A^*$ and $w \in B^*$ with w sufficiently longer than p, can one always find a nonerasing morphism $h\colon A^* \to B^*$ such that $h(p)$ is a factor of w? He answered himself in the negative (see the second example above). The present notion of pattern appeared independently in [BEM1979] and [Zim1982].

The question of Abelian pattern avoidance was first posed by Paul Erdős [Erd1961], hidden among a list of 66 other unsolved research problems. An important early result is due to Frederik Michel Dekking.

Theorem 5.3 (Dekking [Dek1979]). *The pattern a^4 is Abelian 2-avoidable.*

We indicate his argument below. Using a similar argument, Dekking also showed that a^3 is Abelian 3-avoidable. He furthermore raised the question whether a^2 is Abelian 4-avoidable. An early step towards the ultimate answer was provided by A. A. Evdokimov [Evd1968], who gave an example of an infinite word without Abelian squares over 25 letters. This was improved to 5 letters by Peter A. B. Pleasants [Ple1970]. The final answer was given by Veikko Keränen: he gives an 85-uniform morphism over 4 letters generating an infinite word without Abelian squares [Ker1992].

Turning to Dekking's argument, let $\phi\colon \{0,1\}^* \to \{0,1\}^*$ be the morphism taking $(0,1)$ to $(0001, 011)$. We will show that $\phi^\infty(0)$ avoids the pattern a^4 in the Abelian sense. We will also make use of an auxillary morphism $g\colon \{0,1\}^* \to \mathbb{Z}/5\mathbb{Z}$ taking $(0,1)$ to $(2,-1)$.

Lemma 5.4. *Fix a sequence of letters $a_1, \ldots, a_n \in \{0,1\}$ and consider factorizations $\phi(a_i) = p_i s_i$ for some choice of prefixes p_i and suffixes $s_i \neq \epsilon$. If $g(p_1) \equiv \cdots \equiv g(p_n) \bmod 5$, then $p_1 = \cdots = p_n$ or $s_1 = \cdots = s_n$.*

PROOF. Considering the prefixes p of 0001 and 011, the possible values for $g(p)$ are 0, 1, 2 and 4. In particular, $g(0001) \equiv g(011) \equiv 0 \bmod 5$.

Now, if $g(p_i) \equiv 0$ for all i, then $p_i = \epsilon$ for all i. If $g(p_i) \equiv 1$ for all i, then $s_i = 1$ for all i. If $g(p_i) \equiv 2$ for all i, then $p_i = 0$ for all i. If $g(p_i) \equiv 4$ for all i, then $p_i = 00$ for all i. □

Lemma 5.5. *Let $q = q_1 \cdots q_m$ be a pattern, with $q_i \in \{0,1\}$, and suppose that $Q_1 \cdots Q_m$ is a shortest Abelian instance of q in $\phi^\infty(0)$. Then $g(Q_j) \not\equiv 0 \bmod 5$ for some $1 \leq j \leq m$.*

PROOF. Choose a prefix $w = a_1 \cdots a_N$ of $\phi^\infty(0)$ so that $Q_1 \cdots Q_m$ is a factor of $\phi(a_1 \cdots a_N)$. We may assume that $\phi(a_1 \cdots a_N) = PQ_1 \cdots Q_m S$ for some $P, S \in \{0,1\}^*$ with $S \neq \epsilon$ (replacing N by $N+1$ if necessary). See Figure 1. Suppose that $g(Q_j) \equiv 0 \bmod 5$ for all $1 \leq j \leq m$. We will construct a shorter Abelian instance $t_1 \cdots t_m$ of q within $a_1 \cdots a_N$, contradicting the minimality of $|Q_1 \cdots Q_m|$.

We first argue for the existence of a subsequence $i_1 < i_2 \cdots < i_m$ of $\{1, 2, \ldots, N\}$ with the property that $\phi(a_{i_j})$ ends within Q_j. That is, there is a nonempty suffix s_j of $\phi(a_{i_j})$ that is also a prefix of Q_j. See Figure 2. If this is not the case, then there is an index $j \geq 2$ yielding the situation illustrated in Figure 3 (i.e., Q_{j-1} is a proper factor of $\phi(a_{i_j})$). We deduce that $g(p_{j-1}) \equiv g(p_j)$ (since $g(Q_{j-1}) \equiv 0$), and moreover that $p_{j-1} = p_j$ (applying Lemma 5.4 with $n = 2$ and $a_1 = a_2 = a_{i_j}$). But then $Q_{j-1} = \epsilon$, which is absurd.

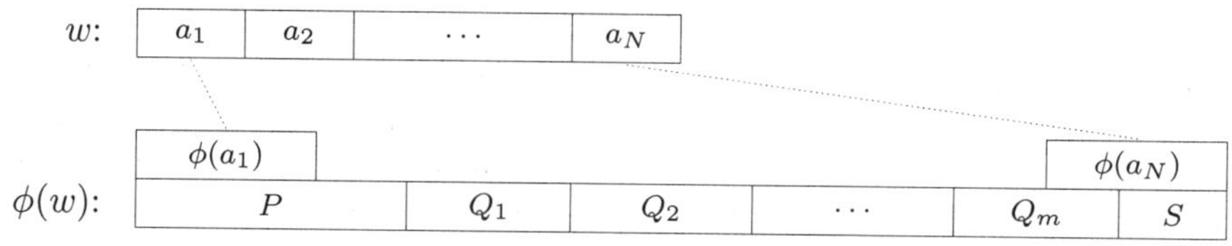

FIGURE 1. If $Q_1 \cdots Q_m$ occurs in $\phi^\infty(0)$, then there is a (minimal) N and a prefix $w = a_1 \cdots a_N$ of $\phi^\infty(0)$ so that $\phi(w)$ has the factorization $PQ_1 \cdots Q_m S$ with $S \neq \epsilon$.

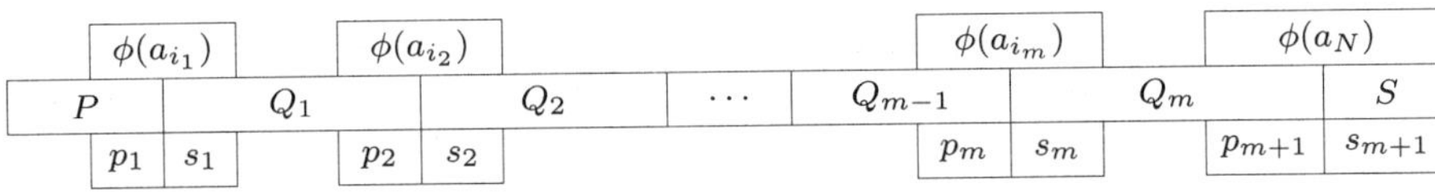

FIGURE 2. Analysis of the prefix $PQ_1 \cdots Q_m S$ of $\phi^\infty(0)$. Given the preimage $a_1 \cdots a_N$, there exist indices $1 \leq i_1 < \cdots < i_m < i_{m+1} = N$ so that $\phi(a_{i_j}) = p_j s_j$ is a factorization centered on $(PQ_1 \cdots Q_{j-1}, Q_j \cdots Q_m S)$ with $s_j \neq \epsilon$.

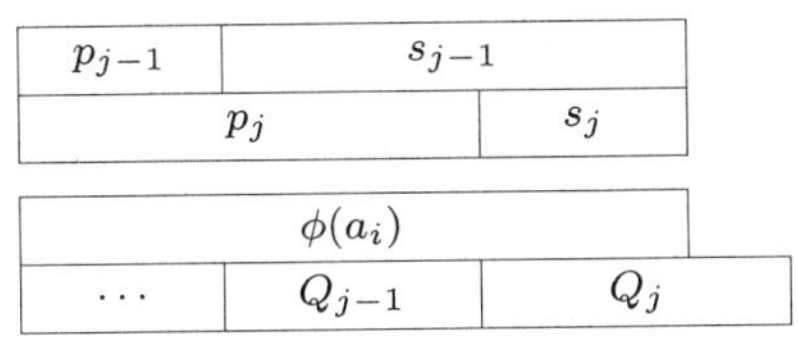

FIGURE 3. Analysis of the prefix $PQ_1 \cdots Q_m S$ of $\phi^\infty(0)$. An impossible relationship between the $\phi(a_i)$s and the Q_js.

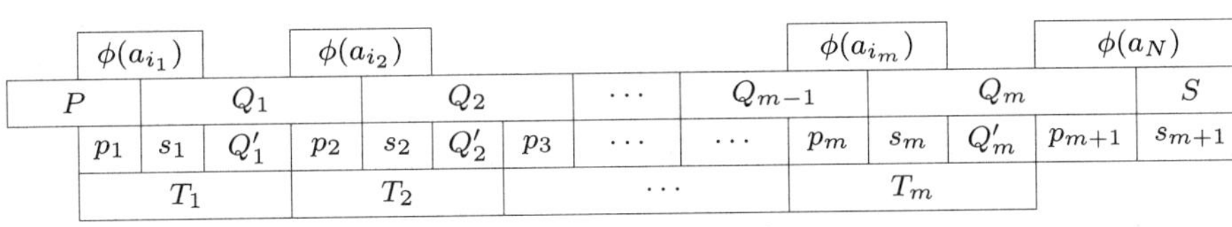

FIGURE 4. Analysis of the prefix $PQ_1 \cdots Q_m S$ of $\phi^\infty(0)$. Factors T_j and Q'_j are added to Figure 2 in order to apply Lemma 5.4.

In summation, we do indeed have the situation illustrated in Figure 2. Let us augment that picture by labelling a few more factors. See Figure 4. Note that each Q'_j is in the image of ϕ, so we get $g(Q'_j) \equiv 0$ (indeed, $g(\phi(0))$ and $g(\phi(1))$ are both zero modulo 5). Since $g(Q_j) \equiv 0$ for all $1 \leq j \leq m$, we must have $g(s_j)+g(p_{j+1}) \equiv 0$ for all $1 \leq j \leq m$. On the other hand, $0 \equiv g(\phi(a_{i_j})) \equiv g(p_j) + g(s_j)$ for all $1 \leq j \leq m$, from which we infer that $g(p_j) \equiv g(p_{j+1}) \bmod 5$ for all $1 \leq j \leq m$. Lemma 5.4 then tells us that $p_1 = \cdots = p_{m+1}$ or $s_1 = \cdots = s_{m+1}$. We consider the first case (the second being symmetric).

Referring to Figure 4, note that T_j is $\phi(t_j)$ for some word t_j, and that $t_1 \cdots t_m$ is a factor of $\phi^\infty(0)$ satisfying $|t_1 \cdots t_m| < |Q_1 \cdots Q_m|$ (because $|\phi(0)|, |\phi(1)| > 1$). We will show that $t_1 \cdots t_m$ is an Abelian instance of q in $\phi^\infty(0)$, contradicting the assumption that $Q_1 \cdots Q_m$ is a shortest Abelian instance of q in $\phi^\infty(0)$. Suppose $q_i = q_j$; we need to show that $|t_i|_z = |t_j|_z$ for $z \in \{0,1\}$. Since $Q_1 \cdots Q_m$ is an Abelian instance of q, we have $Q_i \sim Q_j$. Looking again at Figure 4, we infer that $T_k p_{k+1} = p_k Q_k$ for all $1 \leq k \leq m$. Since $p_k = p_{k+1}$ for all $1 \leq k \leq m$, it follows that $T_k \sim Q_k$ for all $1 \leq k \leq m$. Consequently, $T_i \sim T_j$, so $|T_i|_z = |T_j|_z$ for all $z \in \{0,1\}$. Since $|T_k|_0 = 3|t_k|_0 + |t_k|_1$ and $|T_k|_1 = |t_k|_0 + 2|t_k|_1$ for all $1 \leq k \leq m$, we have

$$3|t_i|_0 + |t_i|_1 = 3|t_j|_0 + |t_j|_1 \quad \text{and} \quad |t_i|_0 + 2|t_i|_1 = |t_j|_0 + 2|t_j|_1.$$

It follows that $|t_i|_z = |t_j|_z$ for $z \in \{0,1\}$. Hence, $t_1 \cdots t_m$ is an Abelian instance of q in $\phi^\infty(0)$. □

Proof of Theorem 5.3. Suppose a^4 is not Abelian 2-avoidable in $\phi^\infty(0)$ and consider a shortest Abelian instance $PQ_1Q_2Q_3Q_4R$ of a^4. Since $Q_i \sim Q_j$ for all $1 \le i, j \le 4$, we have that $g(Q_j)$ is constant for all j (and nonzero by Lemma 5.5). In particular, the sequence

$$g(P), \quad g(PQ_1), \quad g(PQ_1Q_2), \quad g(PQ_1Q_2Q_3), \quad g(PQ_1Q_2Q_3Q_4) \tag{5.6}$$

is an arithmetic progression of length 5. We claim that it does not contain the number 3, which is a contradiction.

To see the claim, build a factorization of P and each Q_j, analogous to that in the proof of Lemma 5.5, as follows: Find the unique i_1 so that

$$|\phi(a_1 \cdots a_{i_1-1})| \le |P| < |\phi(a_1 \cdots a_{i_1})|.$$

Write $P = \phi(a_1 \cdots a_{i_1-1})p_1$ for some prefix p_1 of $\phi(a_{i_1})$. Repeat for each factor in (5.6), writing, e.g., $PQ_1 = \phi(a_1 \cdots a_{i_2-1})p_2$ for some prefix p_2 of $\phi(a_{i_2})$. (We do not demand that $i_1 < i_2 < \cdots < i_5$.) Since $g(t) \equiv 0$ for any t in the image of ϕ, the sequence in (5.6) becomes

$$g(p_1), \quad g(p_2), \quad g(p_3), \quad g(p_4), \quad g(p_5).$$

The prefixes above were chosen so that $\phi(a_{i_j}) = p_j s_j$ with $s_j \neq \epsilon$. We observed in the proof of Lemma 5.4 that $g(p)$ must belong to $\{0, 1, 2, 4\}$ for such prefixes p. This completes the proof of the claim and the theorem. □

Dekking's technique of comparing arithmetic progressions works for more general patterns than just powers. Consider the following example.

Example. The word $\phi^\infty(0)$ avoids $a^3ba^2b^3$ in the Abelian sense (Exercise 5.7).

The above example is a "true" generalization of Dekking's theorem, as opposed to the Abelian 2-avoidability of, say, *abababba*, which is the same as $(ab)^4$ (indeed as a^4) in the sense of Abelian pattern avoidance. James Currie and Terry Visentin have begun to investigate which binary patterns are Abelian 2-avoidable. In [CV2007], many such patterns are found; the avoidance proofs are in the spirit of the Dekking argument above.

Exercise 5.3 (König's Lemma [Lot2002, Proposition 1.2.3]**).** If X is an infinite prefix-closed set of words over a finite alphabet A, then there is an infinite word x having all of its prefixes in X.

Exercise 5.4 ([Lot2002, Proposition 1.6.3])**.** A pattern p is k-avoidable if and only if there are infinitely many words in $\{0, 1, \ldots, k-1\}^*$ that avoid p. (*Hint*: A standard application of König's Lemma above.)

Exercise 5.5 ([Dek1979])**.** Show that a^3 is Abelian 3-avoidable.

Exercise 5.6. The pattern a^2 is not Abelian 3-avoidable. The pattern a^3 is not Abelian 2-avoidable. (*Hint*: Maximal words avoiding the patterns have length seven and nine, respectively.)

Exercise 5.7. The pattern $a^3ba^2b^3$ is Abelian 2-avoidable. (*Hint*: Use the word $\phi^\infty(0)$ and argue, as in the proof of Theorem 5.3, with arithmetic progressions.)

4. Zimin patterns

We conclude our introduction to the theory of patterns with a discussion of certain unavoidable patterns. The reader interested in learning more about unavoidability may consult [Lot2002, Chapter 3]. See also [Cur1993], where a number of open questions are posed — some with prizes attached.

Definition 5.7. Given a fixed alphabet A, the set of **Zimin words** $Z(A)$ are the words in A^* defined recursively as follows:

(i) every letter $a \in A$ is a Zimin word;
(ii) if p is a Zimin word over $A \setminus \{a\}$, then pap is a Zimin word.

Example. The Zimin words $Z(\{a,b,c\})$ include a, aba, $abacaba$ and all images of these under permutations of the ordered alphabet (a,b,c).

As it happens, the Zimin words are unavoidable in a sense we now make precise (indeed, this is why A. I. Zimin introduced them [Zim1982]). A pattern p over A is called **k-unavoidable** if every sufficiently long word w over a k-letter alphabet B contains an instance of p. More precisely, there is an integer N so that if $|w| \geq N$, then there is a nonerasing morphism $h_w \colon A^* \to B^*$ so that $h_w(p)$ is a factor of w.

We will also need an avoidability notion for sets. Given a set $P = \{p_1, p_2, \ldots, p_n\}$ of words in A^*, we say that P is a **k-unavoidable set** if every sufficiently long word w over a k-letter alphabet B has an associated nonerasing morphism $h_w \colon A^* \to B^*$ satisfying $h_w(p_i)$ is a factor of w for all $1 \leq i \leq n$. Conversely, we say that P is a **k-avoidable set** if there is an infinite word $\boldsymbol{x}$ over B with $|B| = k$ so that, no matter the morphism $h \colon A^* \to B^*$, $\boldsymbol{x}$ does not have the factor $h(p_i)$ for at least one $p_i \in P$.

Note that in the case $|P| = 1$, this notion reduces to the preceding pattern-avoidance notion. We call P a set of patterns in what follows to emphasize the connection. Finally, given two sets of patterns P and P', we write $P \doteq P'$ if for each $k \in \mathbb{N}$ either both are k-avoidable or both are k-unavoidable.

Proposition 5.8. *Fix a letter $a \in A$ and two words $p_1, p_2 \in (A \setminus a)^*$. Then*

$$\{p_1, p_2\} \doteq \{p_1 a p_2\}.$$

PROOF. Fix k and suppose that $\{p_1, p_2\}$ is a k-avoidable set. Then there is an infinite word $\boldsymbol{x}$ over $B = \{0, 1, \ldots, k-1\}$ such that, no matter the morphism $h_{\hat{a}} \colon (A \setminus a) \to B$, either $h_{\hat{a}}(p_1)$ or $h_{\hat{a}}(p_2)$ is not a factor of $\boldsymbol{x}$. Now, $\{p_1 a p_2\}$ being a k-unavoidable set would mean that $h(p_1)h(a)h(p_2)$ exists within some prefix of $\boldsymbol{x}$ (for some $h \colon A^* \to B^*$). Defining $h_{\hat{a}}$ to be the restriction of h to $A \setminus a$, this forces both $h_{\hat{a}}(p_1)$ and $h_{\hat{a}}(p_2)$ to be factors of $\boldsymbol{x}$. Consequently, $\{p_1 a p_2\}$ must be a k-avoidable set.

Conversely, suppose that $\{p_1 a p_2\}$ is a k-avoidable set. Then there is an infinite word $\boldsymbol{x}$ over $B = \{0, 1, \ldots, k-1\}$ such that, no matter the morphism $h \colon A^* \to B^*$, $\boldsymbol{x}$ does not contain the factor $h(p_1)h(a)h(p_2)$. Now, $\{p_1, p_2\}$ being a k-unavoidable set would mean that there is an integer N satisfying: for all $w \in B^n$ ($n \geq N$), there exists a nonerasing morphism $h_w \colon (A \setminus a)^* \to B^*$ such that both $h_w(p_1)$ and $h_w(p_2)$ are factors of w. Being an infinite word, $\boldsymbol{x}$ cannot avoid every word of length N. In fact, there must be a word w of length N yielding the factorization

$$\boldsymbol{x} = uwvw\boldsymbol{x}' \quad \text{with } v \neq \epsilon$$

for some $u, v \in B^*$. Writing $w = w_1' h_w(p_1) w_1''$ and $w = w_2' h_w(p_2) w_2''$, we further factorize $\boldsymbol{x}$ as

$$uw_1' h_w(p_1) w_1'' v w_2' h_w(p_2) w_2'' \boldsymbol{x}'.$$

Extending h_w to a morphism h of A^* by defining $h(a) = w_1'' v w_2'$ would then contradict our assumptions on $\{p_1 a p_2\}$. Consequently, $\{p_1, p_2\}$ is a k-avoidable set. □

Corollary 5.9. *The Zimin words are k-unavoidable for every $k \in \mathbb{N} \setminus \{0\}$.*

PROOF. Fix an alphabet A and a Zimin word $w \in Z(A)$. We reason by induction on the length ℓ of w, supposing every Zimin word having less than $|A|$ distinct letters and of length less than ℓ has been shown k-unavoidable for all k.

The base case $\ell = 1$ holds because each $a \in A$ is k-unavoidable for all $k \in \mathbb{N} \setminus \{0\}$. If w is a Zimin word of length $\ell > 1$, then $w = pap$ for some $a \in A$ and $p \in (A \setminus a)^* \setminus \{\epsilon\}$. Moreover, $p \in Z(A) \cap (A \setminus \{a\})^*$ by definition. Proposition 5.8 then gives

$$\{pap\} \doteq \{p, p\} \doteq \{p\}.$$

Now, p is k-unavoidable for all k by induction, implying pap is as well. □

In a certain sense, Zimin words are the *only* unavoidable patterns. This is indicated by the next result.

Theorem 5.10 (Zimin, [Zim1982]). *A pattern p on an alphabet A is unavoidable if and only if there exists a Zimin word z on an alphabet B and a nonerasing morphism $h \colon A^* \to B^*$ so that $h(p)$ appears as a factor of z.*

Example. The pattern $p = abcba$ is unavoidable because of the Zimin word $z = abacaba$. More specifically, the morphism h defined by $(a, b, c) \mapsto (b, a, c)$ maps p to $bacab$, the central factor of z.

Finding the morphism h in the theorem may be difficult. Luckily, Proposition 5.8 can give us more than Corollary 5.9. To illustrate, consider the pattern $p = abxbcycazactcb$ on the alphabet $A = \{a, b, c, x, y, z, t\}$.

Corollary 5.11. *The pattern p above is 3-unavoidable.*

PROOF. Using the proposition, we have

$$\{p\} \doteq \{abxbcyca, actcb\}$$

(eliminating z). Continuing by eliminating y, x and t, we arrive at

$$\{p\} \doteq \{ab, bc, ca, ac, cb\}.$$

We show that $Q = \{ab, bc, ca, ac, cb\}$ is a 3-unavoidable set by considering the words w in $\{0, 1, 2\}^*$ of length 14. If w contains a square, then w does not avoid Q (take $h(a) = h(b) = h(c) = u$). Otherwise, one checks that w contains all six factors ij with $\{i, j\} \subseteq \{0, 1, 2\}$ (there are many square-free words of length 14, but if Thue could do it, so can you). Taking $h(a) = 0$, $h(b) = 1$ and $h(c) = 2$, we see that such a w also does not avoid Q. Finally, if w is a word of length 15 or greater, then it has a prefix of length 14, so it also does not avoid Q. This completes the proof. □

Exercise 5.8. If a pattern p is $(k + 1)$-unavoidable for some k, then it is also k-unavoidable (and the same bound N on the length of exceptional words will suffice).

Exercise 5.9. A set of patterns P is k-unavoidable if and only if it is not k-avoidable.

Exercise 5.10 (Zimin images). In Theorem 5.10, it was shown that morphic preimages of factors of Zimin words are unavoidable. Show that the same does not hold for morphic images. Consider aba and find a nonerasing endomorphism γ on $\{a,b\}^*$ so that $\gamma(aba)$ is 2-avoidable.

5. Bi-ideal sequences

The recursive form of Zimin words has been exploited in many ways not outlined above. We mention a few of them here and refer the reader to [Lot2002, Chapter 4] for more details.

Zimin words have a natural generalization called **sesquipowers**. These are defined recursively as follows: any nonempty word is a sesquipower of order 1; a word w over an alphabet A is a sesquipower of order $n > 1$ if $w = w_0 v w_0$ for some words $w_0, v \in A^*$ with $w_0 \neq \epsilon$ and w_0 a sesquipower of order $n-1$. So w is a sesquipower if and only if it is a nonempty image of a Zimin word under a morphism h (not necessarily nonerasing). A sequence of words $w_n \in A^*$ is called a **bi-ideal sequence** if each term in the sequence is a sesquipower of the preceding term, i.e.,

$$w_n = w_{n-1} v_n w_{n-1}$$

for some choice of words $v_n \in A^*$.

The use of bi-ideal sequences in algebraic structures is quite old (see, e.g., *m-sequences* in [Jac1956] or [Sch1961, Section IV.5]), but the terminology was not coined until 1966 by Michel Coudrain and Marcel-Paul Schützenberger. In [CS1966], they use bi-ideal sequences to give criteria for a finitely generated semigroup to be finite (see [Lot2002, Theorem 4.5.10]).

Closer to the topics in the present book, bi-ideal sequences may also be used to improve upon a classical result of Anatoly I. Shirshov [Šir1957] on *n-divisions*. Suppose u is a word over a totally-ordered alphabet. A factorization $u = (x_1, x_2, \ldots, x_n)$ is called an ***n*-division** if each x_i is nonempty and u is lexicographically greater than any nontrivial anagram $x_{i_1} x_{i_2} \cdots x_{i_n}$ of the factors. The following appears as Theorem 4.4.5 in [Lot2002].

Theorem 5.12 (de Luca, Varricchio [dLV1999]). *Fix a totally-ordered alphabet A of cardinality k. Given positive integers p and n, every sufficiently long word $w \in A^*$ satisfies*:

(i) *there exists $u \neq \epsilon$ such that u^p is a factor of w; or*

(ii) *there exists a factor u of w that is the nth term of a bi-ideal sequence. Moreover, u has an n-division $u = (x_1, x_2, \ldots, x_n)$ where each x_i is a Lyndon word and $x_1 > x_2 > \cdots > x_n$.*

Finally, to any bi-ideal sequence $(w_n)_{n \geq 1}$, one may naturally associate an infinite word $\boldsymbol{w} = \lim_{n \to \infty} w_n$ with arbitrarily long sesquipower prefixes. Infinite words constructed in this fashion are precisely the recurrent words. See Exercise 5.11.

Exercise 5.11 ([Lot2002, Proposition 4.3.1])**.** Recall that an infinite word $\boldsymbol{x}$ is **recurrent** if every finite factor of $\boldsymbol{x}$ occurs within $\boldsymbol{x}$ infinitely often. Prove that every recurrent infinite word is the limit of some bi-ideal sequence. (*Hint*: Given a recurrent word $\boldsymbol{x}$, set $w_1 = x_1$ and define a bi-ideal sequence $w_1, w_2, w_3, \ldots$ recursively, using the recurrent property of $\boldsymbol{x}$.)

6. Repetitions in Sturmian words

To tie the two parts of this book together, we briefly mention some results concerning repetitions in infinite words that are constructed from lines of irrational slope as Christoffel words were constructed from line segments of rational slope in Part I.

We begin with the Fibonacci word. Let ϕ be the golden ratio. Let ℓ be the positive ray in $\mathbb{R}^2$ of slope $-\phi^\vee = \phi - 1$ beginning at the origin, and let $\boldsymbol{s}$ be the infinite binary word obtained by discretizing ℓ. The **Fibonacci word**, denoted by $\boldsymbol{f}$, is the infinite binary word satisfying $\boldsymbol{s} = x\boldsymbol{f}$. We have already encountered $\boldsymbol{f}$: it was introduced in Exercise 1.5 of Part I, and studied in Exercises 3.7 and 4.12 of Part II.

It is known from the work of Filippo Mignosi and Giuseppe Pirillo [MP1992] that $\boldsymbol{f}$ contains no powers of exponent greater than $2 + \phi$, and that it contains powers of exponent less than but arbitrarily close to this number. Therefore, the critical exponent of $\boldsymbol{f}$ is $2 + \phi$.

This result has been extended as follows. For any irrational real number α, let $\boldsymbol{s}$ be the infinite word obtained by discretizing the line of slope $\alpha/(1 + \alpha)$ passing through the origin in $\mathbb{R}^2$, and let $\boldsymbol{c}_\alpha$ denote the infinite binary word satisfying $\boldsymbol{s} = x\boldsymbol{c}_\alpha$. Such words are called **characteristic Sturmian words**. Mignosi proved that if the continued fraction representation of α has bounded partial quotients, then the powers occurring in $\boldsymbol{c}_\alpha$ are bounded, and conversely [Mig1991]. Detailed studies of the exact exponent of powers that appear in a Sturmian word have been carried out later.

Finally, we mention a recent, interesting connection between repetitions and transcendence. It has been shown that if a binary infinite word has infinitely many prefixes that are repetitions of exponent $2 + \varepsilon$ for some ε, then the real number whose binary expansion is this infinite word is either rational or transcendental [FM1997]. As a consequence, any number whose binary expansion is a Sturmian word is either rational or transcendental. For more results in this direction, see [AB2005].

Bibliography

For the convenience of the reader, each entry below is followed by ↑ and a list of page numbers indicating the location within the book where the citation occurs.

[ABG2007] A. Aberkane, S. Brlek, and A. Glen, *Sequences having the Thue–Morse complexity* (2007), preprint. ↑84

[AB2005] B. Adamczewski and Y. Bugeaud, *On the decimal expansion of algebraic numbers*, Fiz. Mat. Fak. Moksl. Semin. Darb. **8** (2005), 5–13. ↑134

[AL1977] A. Adler and S.-Y. R. Li, *Magic cubes and Prouhet sequences*, Amer. Math. Monthly **84** (1977), no. 8, 618–627. ↑74

[AAB+1995] J.-P. Allouche, A. Arnold, J. Berstel, S. Brlek, W. Jockusch, S. Plouffe, and B. E. Sagan, *A relative of the Thue–Morse sequence*, Discrete Math. **139** (1995), no. 1-3, 455–461. ↑71, 95

[AARS1994] J.-P. Allouche, D. Astoorian, J. Randall, and J. Shallit, *Morphisms, squarefree strings, and the Tower of Hanoi puzzle*, Amer. Math. Monthly **101** (1994), no. 7, 651–658. ↑90, 95, 99

[AB1992] J.-P. Allouche and R. Bacher, *Toeplitz sequences, paperfolding, Towers of Hanoi and progression-free sequences of integers*, Enseign. Math. (2) **38** (1992), no. 3-4, 315–327. ↑95

[AS1999] J.-P. Allouche and J. Shallit, *The ubiquitous Prouhet–Thue–Morse sequence*, Sequences and Their Applications (Singapore, 1998) (C. Ding, T. Helleseth, and H. Niederreiter, eds.), Springer Ser. Discrete Math. Theor. Comput. Sci., Springer, London, 1999, pp. 1–16. ↑69

[AS2003] ———, *Automatic sequences: Theory, applications, generalizations*, Cambridge Univ. Press, Cambridge, 2003. ↑6, 71, 78, 79, 80, 81

[AT2007] A. Alpers and R. Tijdeman, *The two-dimensional Prouhet–Tarry–Escott problem*, J. Number Theory **123** (2007), no. 2, 403–412. ↑73

[Aus2006] P. Auster, *The Brooklyn follies*, Henry Holt and Co., New York, NY, 2006. ↑45

[BEM1979] D. R. Bean, A. Ehrenfeucht, and G. F. McNulty, *Avoidable patterns in strings of symbols*, Pacific J. Math. **85** (1979), no. 2, 261–294. ↑102, 127

[Ber1771] J. Bernoulli, *Sur une nouvelle espèce de calcul*, Recueil pour les astronomes, Vol. 1, Berlin, 1771, pp. 255–284. ↑3, 55

[Ber1990] J. Berstel, *Tracé de droites, fractions continues et morphismes itérés*, Mots, Lang. Raison. Calc., Hermès, Paris, 1990, pp. 298–309. ↑5

[Ber1995] ———, *Axel Thue's papers on repetitions in words: A translation*, Publications du LaCIM, vol. 20, UQÀM, Montréal, QC, 1995. ↑81

[BdL1997] J. Berstel and A. de Luca, *Sturmian words, Lyndon words and trees*, Theoret. Comput. Sci. **178** (1997), no. 1-2, 171–203. ↑20, 28, 41, 47

[BP1985] J. Berstel and D. Perrin, *Theory of codes*, Pure Appl. Math., vol. 117, Academic Press, Orlando, FL, 1985. ↑105

[BS2006] J. Berstel and A. Savelli, *Crochemore factorization of Sturmian and other infinite words*, Mathematical Foundations of Computer Science 2006 (Stará Lesná, 2006) (R. Královič and P. Urzyczyn, eds.), Lecture Notes in Comput. Sci., vol. 4162, Springer, Berlin, 2006, pp. 157–166. ↑116

[BS1993] J. Berstel and P. Séébold, *A characterization of overlap-free morphisms*, Discrete Appl. Math. **46** (1993), no. 3, 275–281. ↑81

[BdLR2008] V. Berthé, A. de Luca, and C. Reutenauer, *On an involution of Christoffel words and Sturmian morphisms*, European J. Combin. **29** (2008), no. 2, 535–553. ↑14, 15, 25

[BBGL2007] A. Blondin-Massé, S. Brlek, A. Glen, and S. Labbé, *On the critical exponent of generalized Thue–Morse words*, Discrete Math. Theor. Comput. Sci. **9** (2007), no. 1, 293–304. ↑126

[BL1993] J.-P. Borel and F. Laubie, *Quelques mots sur la droite projective réelle*, J. Théor. Nombres Bordeaux **5** (1993), no. 1, 23–51. ↑5, 17, 18, 20, 25, 47

[BR2006] J.-P. Borel and C. Reutenauer, *On Christoffel classes*, Theor. Inform. Appl. **40** (2006), no. 1, 15–27. ↑27, 42, 44

[BI1994] P. Borwein and C. Ingalls, *The Prouhet–Tarry–Escott problem revisited*, Enseign. Math. (2) **40** (1994), no. 1-2, 3–27. ↑73

[BLP2003] P. Borwein, P. Lisoněk, and C. Percival, *Computational investigations of the Prouhet–Tarry–Escott problem*, Math. Comp. **72** (2003), no. 244, 2063–2070. ↑73

[Brl1989] S. Brlek, *Enumeration of factors in the Thue–Morse word*, Discrete Appl. Math. **24** (1989), no. 1-3, 83–96. ↑81, 82

[BW1994] M. Burrows and D. J. Wheeler, *A block-sorting lossless data compression algorithm*, Technical Report 124, Digital, Palo Alto, CA, 1994. ↑39

[Car1921] F. Carlson, *Über Potenzreihen mit ganzzahligen Koeffizienten*, Math. Z. **9** (1921), no. 1-2, 1–13. ↑89

[Car2006] A. Carpi, *On the repetition threshold for large alphabets*, Mathematical Foundations of Computer Science 2006 (Stará Lesná, 2006) (R. Královič and P. Urzyczyn, eds.), Lecture Notes in Comput. Sci., vol. 4162, Springer, Berlin, 2006, pp. 226–237. ↑126

[Car2007] ———, *On Dejean's conjecture over large alphabets*, Theoret. Comput. Sci. **385** (2007), no. 1-3, 137–151. ↑126

[Cas1957] J. W. S. Cassels, *An introduction to Diophantine approximation*, Cambridge Tracts in Mathematics and Mathematical Physics, vol. 45, Cambridge Univ. Press, New York, NY, 1957. ↑56

[CFL1958] K.-T. Chen, R. H. Fox, and R. C. Lyndon, *Free differential calculus*. IV: *The quotient groups of the lower central series*, Ann. of Math. (2) **68** (1958), 81–95. ↑41

[CS1963] N. Chomsky and M.-P. Schützenberger, *The algebraic theory of context-free languages*, Computer Programming and Formal Systems, North-Holland, Amsterdam, 1963, pp. 118–161. ↑89

[Chr1875] E. B. Christoffel, *Observatio arithmetica*, Ann. Mat. Pura Appl. (2) **6** (1875), 148–152. ↑5, 7, 44

[Chr1887] ———, *Lehrsätze über arithmetische Eigenschaften der Irrationalzahlen*, Ann. Mat. Pura Appl. (2) **15** (1887–1888), 253–276. ↑5

[CKMR1980] G. Christol, T. Kamae, M. Mendès France, and G. Rauzy, *Suites algébriques, automates et substitutions*, Bull. Soc. Math. France **108** (1980), no. 4, 401–419. ↑80

[Cob1972] A. Cobham, *Uniform tag sequences*, Math. Systems Theory **6** (1972), 164–192. ↑78

[Coh1972] H. Cohn, *Markoff forms and primitive words*, Math. Ann. **196** (1972), 8–22. ↑11

[CS1966] M. Coudrain and M.-P. Schützenberger, *Une condition de finitude des monoïdes finiment engendrés*, C. R. Acad. Sci. Paris Sér. A-B **262** (1966), A1149–A1151. ↑133

[CH1973] E. M. Coven and G. A. Hedlund, *Sequences with minimal block growth*, Math. Systems Theory **7** (1973), 138–153. ↑43, 85

[Cro1981] M. Crochemore, *An optimal algorithm for computing the repetitions in a word*, Inform. Process. Lett. **12** (1981), no. 5, 244–250. ↑109

[Cro1982] ———, *Sharp characterizations of squarefree morphisms*, Theoret. Comput. Sci. **18** (1982), no. 2, 221–226. ↑100, 101

[Cro1983a] ———, *Recherche linéaire d'un carré dans un mot*, C. R. Acad. Sci. Paris Sér. I Math. **296** (1983), no. 18, 781–784. ↑114, 115

[Cro1983b] ———, *Régularités évitables*, Thèse d'État, Université de Rouen, 1983. ↑100

[Cro1986] ———, *Transducers and repetitions*, Theoret. Comput. Sci. **45** (1986), no. 1, 63–86. ↑122

[CDP2005] M. Crochemore, J. Désarménien, and D. Perrin, *A note on the Burrows – Wheeler transformation*, Theoret. Comput. Sci. **332** (2005), no. 1-3, 567 – 572. ↑39, 40

[CHL2001] M. Crochemore, C. Hancart, and T. Lecroq, *Algorithmique du texte*, Vuibert, Paris, 2001; English transl. in *Algorithms on strings*, Cambridge Univ. Press, Cambridge, 2007. ↑111

[CHL2007] ———, *Algorithms on strings*, Cambridge Univ. Press, Cambridge, 2007. ↑111, 114, 116, 119, 123

[CI2007] M. Crochemore and L. Ilie, *Analysis of maximal repetitions in strings*, Mathematical Foundations of Computer Science 2007 (Ceský Krumlov, 2007) (L. Kucera and A. Kucera, eds.), Lecture Notes in Comput. Sci., vol. 4708, Springer, Berlin, 2007, pp. 465 – 476. ↑125, 126

[CR1995] M. Crochemore and W. Rytter, *Squares, cubes, and time-space efficient string searching*, Algorithmica **13** (1995), no. 5, 405 – 425. ↑107, 109

[CR2002] ———, *Jewels of stringology*, World Sci. Publ.., River Edge, NJ, 2002. ↑117, 124

[Cur1993] J. D. Currie, *Unsolved problems: Open problems in pattern avoidance*, Amer. Math. Monthly **100** (1993), no. 8, 790 – 793. ↑131

[CM2007] J. D. Currie and M. Mohammad-Noori, *Dejean's conjecture and Sturmian words*, European J. Combin. **28** (2007), no. 3, 876 – 890. ↑126

[CV2007] J. D. Currie and T. I. Visentin, *On abelian 2-avoidable binary patterns*, Acta Inform. **43** (2007), no. 8, 521 – 533. ↑130

[CF1989] T. W. Cusick and M. E. Flahive, *The Markoff and Lagrange spectra*, Math. Surveys Monogr., vol. 30, Amer. Math. Soc., Providence, RI, 1989. ↑56, 64

[Dav1992] H. Davenport, *The higher arithmetic: An introduction to the theory of numbers*, 6th ed., Cambridge Univ. Press, Cambridge, 1992. ↑47

[Dej1972] F. Dejean, *Sur un théorème de Thue*, J. Combinatorial Theory Ser. A **13** (1972), 90 – 99. ↑126

[Dek1979] F. M. Dekking, *Strongly nonrepetitive sequences and progression-free sets*, J. Combin. Theory Ser. A **27** (1979), no. 2, 181 – 185. ↑128, 130

[Dic1930] L. E. Dickson, *Studies in the theory of numbers*, Univ. Chicago Press, Chicago, IL, 1930; reprinted as *Introduction to the theory of numbers*, Dover Publ., New York, NY, 1957. ↑55, 56

[DB1937] H. L. Dorwart and O. E. Brown, *The Tarry – Escott problem*, Amer. Math. Monthly **44** (1937), no. 10, 613 – 626. ↑71

[DG1990] S. Dulucq and D. Gouyou-Beauchamps, *Sur les facteurs des suites de Sturm*, Theoret. Comput. Sci. **71** (1990), no. 3, 381 – 400. ↑41

[Duv1983] J.-P. Duval, *Factorizing words over an ordered alphabet*, J. Algorithms **4** (1983), no. 4, 363 – 381. ↑41

[Erd1961] P. Erdős, *Some unsolved problems*, Magyar Tud. Akad. Mat. Kutató Int. Közl. **6** (1961), 221 – 254. ↑127

[Evd1968] A. A. Evdokimov, *Strongly asymmetric sequences generated by a finite number of symbols*, Dokl. Akad. Nauk SSSR **179** (1968), 1268 – 1271; English transl., Soviet Math. Dokl. **9** (1968), 536 – 539. ↑128

[FM1997] S. Ferenczi and C. Mauduit, *Transcendence of numbers with a low complexity expansion*, J. Number Theory **67** (1997), no. 2, 146 – 161. ↑134

[FS1998] A. S. Fraenkel and J. Simpson, *How many squares can a string contain?*, J. Combin. Theory Ser. A **82** (1998), no. 1, 112 – 120. ↑109, 110

[Fro1913] F. G. Frobenius, *Über die Markoffschen Zahlen*, Sitzungsberichte der Königlich Preussischen Akademie der Wissenschaften zu Berlin (1913), 458 – 487; reprinted in J.-P. Serre (ed.), *Gesammelte Abhandlungen*, Vol. 3, Springer, Berlin, 1968. ↑56, 58

[Fur1967] H. Furstenberg, *Algebraic functions over finite fields*, J. Algebra **7** (1967), 271 – 277. ↑80

[Gau1986] C. F. Gauss, *Disquisitiones arithmeticae*, Springer, New York, NY, 1986. ↑55

[GR1993] I. M. Gessel and C. Reutenauer, *Counting permutations with given cycle structure and descent set*, J. Combin. Theory Ser. A **64** (1993), no. 2, 189 – 215. ↑39

[GLS2008] A. Glen, A. Lauve, and F. V. Saliola, *A note on the Markoff condition and central words*, Inform. Process. Lett. **105** (2008), no. 6, 241 – 244. ↑61, 64

[GV1991] P. Goralčík and T. Vanicek, *Binary patterns in binary words*, Internat. J. Algebra Comput. **1** (1991), no. 3, 387 – 391. ↑105

[GKP1994] R. L. Graham, D. E. Knuth, and O. Patashnik, *Concrete mathematics: A foundation for computer science*, 2nd ed., Addison-Wesley, Reading, MA, 1994. ↑47, 53

[Gus1997] D. Gusfield, *Algorithms on strings, trees, and sequences: Computer science and computational biology*, Cambridge Univ. Press, Cambridge, 1997. ↑117, 119, 122

[Hal1964] M. Hall Jr., *Generators and relations in groups: The Burnside problem*, Lectures on Modern Mathematics, Vol. II, Wiley, New York, NY, 1964, pp. 42–92. ↑98, 99

[HYY2003] H. K. Hsiao, Y. T. Yeh, and S. S. Yu, *Square-free-preserving and primitive-preserving homomorphisms*, Acta Math. Hungar. **101** (2003), no. 1-2, 113–130. ↑101

[Ili2007] L. Ilie, *A note on the number of squares in a word*, Theoret. Comput. Sci. **380** (2007), no. 3, 373–376. ↑110

[Jac1956] N. Jacobson, *Structure of rings*, Amer. Math. Soc., Colloq. Publ., vol. 37, Amer. Math. Soc., Providence, RI, 1956. ↑133

[JZ2004] O. Jenkinson and L. Q. Zamboni, *Characterisations of balanced words via orderings*, Theoret. Comput. Sci. **310** (2004), no. 1-3, 247–271. ↑43

[Jus2005] J. Justin, *Episturmian morphisms and a Galois theorem on continued fractions*, Theor. Inform. Appl. **39** (2005), no. 1, 207–215. ↑25, 29

[KR2007] C. Kassel and C. Reutenauer, *Sturmian morphisms, the braid group* B_4*, Christoffel words and bases of* F_2, Ann. Mat. Pura Appl. (4) **186** (2007), no. 2, 317–339. ↑34

[Ker1986] V. Keränen, *On the k-freeness of morphisms on free monoids*, Ann. Acad. Sci. Fenn. Math. Diss. (1986), no. 61, 55. ↑105

[Ker1987] ———, *On the k-freeness of morphisms on free monoids*, STACS 87 (Passau, 1987) (F.-J. Brandenburg, G. Vidal-Naquet, and M. Wirsing, eds.), Lecture Notes in Comput. Sci., vol. 247, Springer, Berlin, 1987, pp. 180–188. ↑105

[Ker1992] ———, *Abelian squares are avoidable on 4 letters*, Automata, Languages and Programming (Vienna, 1992) (W. Kuich, ed.), Lecture Notes in Comput. Sci., vol. 623, Springer, Berlin, 1992, pp. 41–52. ↑128

[KK1999a] R. Kolpakov and G. Kucherov, *Finding maximal repetitions in a word in linear time*, 40th Annual Symposium on Foundations of Computer Science (New York, NY, 1999), IEEE Computer Soc., Los Alamitos, CA, 1999, pp. 596–604. ↑125

[KK1999b] ———, *On maximal repetitions in words*, Fundamentals of Computation Theory (Iaşi, 1999) (G. Ciobanu and G. Păun, eds.), Lecture Notes in Comput. Sci., vol. 1684, Springer, Berlin, 1999, pp. 374–385. ↑125

[KZ1873] A. Korkine and G. Zolotareff, *Sur les formes quadratiques*, Math. Ann. **6** (1873), no. 3, 366–389. ↑55

[LS1993] G. M. Landau and J. P. Schmidt, *An algorithm for approximate tandem repeats*, Combinatorial Pattern Matching (Padova, 1993) (A. Apostolico, M. Crochemore, Z. Galil, and U. Manber, eds.), Lecture Notes in Comput. Sci., vol. 684, Springer, Berlin, 1993, pp. 120–133. ↑123

[LV1986] G. M. Landau and U. Vishkin, *Introducing efficient parallelism into approximate string matching and a new serial algorithm*, Proc. 18th Annual ACM Symposium on Theory of Computing (Berkeley, CA, 1986), ACM Press, New York, NY, 1986, pp. 220–230. ↑123

[Lec1985] M. Leconte, *A characterization of power-free morphisms*, Theoret. Comput. Sci. **38** (1985), no. 1, 117–122. ↑105

[Lem1970] A. Lempel, *On a homomorphism of the de Bruijn graph and its applications to the design of feedback shift registers*, IEEE Trans. Computers **C-19** (1970), 1204–1209. ↑78

[LZ1976] A. Lempel and J. Ziv, *On the complexity of finite sequences*, IEEE Trans. Information Theory **IT-22** (1976), no. 1, 75–81. ↑114

[Lot1997] M. Lothaire, *Combinatorics on words*, Cambridge Math. Lib., Cambridge Univ. Press, Cambridge, 1997. ↑41, 99, 125

[Lot2002] ———, *Algebraic combinatorics on words*, Encyclopedia Math. Appl., vol. 90, Cambridge Univ. Press, Cambridge, 2002. ↑6, 9, 14, 41, 43, 125, 130, 131, 133

[Lot2005] ———, *Applied combinatorics on words*, Encyclopedia Math. Appl., vol. 105, Cambridge Univ. Press, Cambridge, 2005. ↑110

[dL1997] A. de Luca, *Sturmian words: Structure, combinatorics, and their arithmetics*, Theoret. Comput. Sci. **183** (1997), no. 1, 45–82. ↑24, 25, 27 [dLM1994] A. de Luca and F. Mignosi, *Some combinatorial properties of Sturmian words*, Theoret. Comput. Sci. **136** (1994), no. 2, 361–385. ↑11, 31, 41

[dLV1989] A. de Luca and S. Varricchio, *Some combinatorial properties of the Thue–Morse sequence and a problem in semigroups*, Theoret. Comput. Sci. **63** (1989), no. 3, 333–348. ↑82

[dLV1999] ———, *Finiteness and regularity in semigroups and formal languages*, Monogr. Theoret. Comput. Sci. EATCS Ser., Springer, Berlin, 1999. ↑133

[Luc1893] É. Lucas, *Récréations mathématiques,* Vol. 3, Gauthier-Villars, Paris, 1893; reprinted by A. Blanchard, Paris, 1979. ↑91

[LS2001] R. C. Lyndon and P. E. Schupp, *Combinatorial group theory*, Classics Math., Springer, Berlin, 2001. ↑35

[MKS2004] W. Magnus, A. Karrass, and D. Solitar, *Combinatorial group theory: Presentations of groups in terms of generators and relations*, 2nd ed., Dover Publ., Mineola, NY, 2004. ↑35

[MRS2003] S. Mantaci, A. Restivo, and M. Sciortino, *Burrows–Wheeler transform and Sturmian words*, Inform. Process. Lett. **86** (2003), no. 5, 241–246. ↑40

[Man2001] G. Manzini, *An analysis of the Burrows–Wheeler transform*, J. ACM **48** (2001), no. 3, 407–430. ↑39

[Mar1879] A. Markoff, *Sur les formes quadratiques binaires indéfinies*, Math. Ann. **15** (1879), no. 3-4, 381–406. ↑5, 55, 56, 57

[Mar1880] ———, *Sur les formes quadratiques binaires indéfinies.* 2nd mémoire, Math. Ann. **17** (1880), no. 3, 379–399. ↑5, 55, 56, 57

[Mar1881] ———, *Sur une question de Jean Bernoulli*, Math. Ann. **19** (1881), no. 1, 27–36. ↑5, 55

[McC1976] E. M. McCreight, *A space-economical suffix tree construction algorithm*, J. Assoc. Comput. Mach. **23** (1976), no. 2, 262–272. ↑119

[Mel1999] G. Melançon, *Visualisation de graphes et combinatoire des mots*, thèse d'habilitation à diriger les recherches, Université Bordeaux I, 1999. ↑42

[Men2003] F. Mendivil, *Fractals, graphs, and fields*, Amer. Math. Monthly **110** (2003), no. 6, 503–515. ↑78

[Mig1991] F. Mignosi, *On the number of factors of Sturmian words*, Theoret. Comput. Sci. **82** (1991), no. 1, 71–84. ↑134

[MP1992] F. Mignosi and G. Pirillo, *Repetitions in the Fibonacci infinite word*, RAIRO Inform. Théor. Appl. **26** (1992), no. 3, 199–204. ↑134

[MS1993] F. Mignosi and P. Séébold, *Morphismes sturmiens et règles de Rauzy*, J. Théor. Nombres Bordeaux **5** (1993), no. 2, 221–233. ↑14

[Mor2004] E. Moreno, *On the theorem of Fredricksen and Maiorana about de Bruijn sequences*, Adv. in Appl. Math. **33** (2004), no. 2, 413–415. ↑78

[MH1940] M. Morse and G. A. Hedlund, *Symbolic dynamics.* II: *Sturmian trajectories*, Amer. J. Math. **62** (1940), 1–42. ↑43, 44

[MH1944] ———, *Unending chess, symbolic dynamics and a problem in semigroups*, Duke Math. J. **11** (1944), 1–7. ↑97, 98, 99

[Mou1992] J. Moulin-Ollagnier, *Proof of Dejean's conjecture for alphabets with* $5, 6, 7, 8, 9, 10$ *and* 11 *letters*, Theoret. Comput. Sci. **95** (1992), no. 2, 187–205. ↑126

[Mye1986] E. W. Myers, *An* $O(ND)$ *difference algorithm and its variations*, Algorithmica **1** (1986), no. 1, 251–266. ↑123

[NBT1963] H. Noland, C. H. Braunholtz, and T. Tamura, *Advanced problems and solutions: Solutions*: 5030, Amer. Math. Monthly **70** (1963), no. 6, 675–676. ↑97

[OZ1981] R. P. Osborne and H. Zieschang, *Primitives in the free group on two generators*, Invent. Math. **63** (1981), no. 1, 17–24. ↑5, 34

[Pan1984] J.-J. Pansiot, *À propos d'une conjecture de F. Dejean sur les répétitions dans les mots*, Discrete Appl. Math. **7** (1984), no. 3, 297–311. ↑126

[Pir1999] G. Pirillo, *A new characteristic property of the palindrome prefixes of a standard Sturmian word*, Sém. Lothar. Combin. **43** (1999), B43f. ↑39

[Ple1970] P. A. B. Pleasants, *Non-repetitive sequences*, Proc. Cambridge Philos. Soc. **68** (1970), 267–274. ↑128

[Pro1851] E. Prouhet, *Mémoire sur quelques relations entre les puissances des nombres*, C. R. Acad. Sci. Paris Sér. I Math. **33** (1851), 225. ↑71, 72

[Pyt2002] N. Pytheas Fogg, *Substitutions in dynamics, arithmetics and combinatorics* (V. Berthé, S. Ferenczi, C. Mauduit, and A. Siegel, eds.), Lecture Notes in Math., vol. 1794, Springer, Berlin, 2002. ↑6

[Ram2007] N. Rampersad, *On the context-freeness of the set of words containing overlaps*, Inform. Process. Lett. **102** (2007), no. 2-3, 74–78. ↑89, 90

[Ran1973] G. N. Raney, *On continued fractions and finite automata*, Math. Ann. **206** (1973), 265–283. ↑28

[Rau1984] G. Rauzy, *Des mots en arithmétique*, Journées d'Avignon. Théorie des langages et complexité des algorithmes (Avignon, 1983) (B. Rozoy, ed.), Publ. Dép. Math. Nouvelle Sér. B, vol. 84-6, Univ. Claude-Bernard, Lyon, 1984, pp. 103–113. ↑20

[Reu2006a] C. Reutenauer, *Mots de Lyndon généralisés*, Sém. Lothar. Combin. **54** (2006), B54h. ↑56

[Reu2006b] ———, *On Markoff's property and Sturmian words*, Math. Ann. **336** (2006), no. 1, 1–12. ↑56, 60, 61, 62, 64

[Ric2003] G. Richomme, *Some non-finitely generated monoids of repetition-free endomorphisms*, Inform. Process. Lett. **85** (2003), no. 2, 61–66. ↑100

[RW2002] G. Richomme and F. Wlazinski, *Some results on k-power-free morphisms*, Theoret. Comput. Sci. **273** (2002), no. 1-2, 119–142. ↑100

[RW2004] ———, *Overlap-free morphisms and finite test-sets*, Discrete Appl. Math. **143** (2004), no. 1-3, 92–109. ↑101

[RW2007] ———, *Existence of finite test-sets for k-power-freeness of uniform morphisms*, Discrete Appl. Math. **155** (2007), no. 15, 2001–2016. ↑101

[Rob2005] W. Robertson, *Square cells: An array cooking lesson*, The PracTEX Journal (2005), no. 2, available at `http://tug.org/pracjourn/2005-2/robertson/`. ↑x

[Ryt2006] W. Rytter, *The number of runs in a string: Improved analysis of the linear upper bound*, STACS 2006 (Marseille, 2006) (B. Durand and W. Thomas, eds.), Lecture Notes in Comput. Sci., vol. 3884, Springer, Berlin, 2006, pp. 184–195. ↑125

[Ryt2007] ———, *The number of runs in a string*, Inform. and Comput. **205** (2007), no. 9, 1459–1469. ↑125

[Sch1961] M.-P. Schützenberger, *On a special class of recurrent events*, Ann. Math. Statist. **32** (1961), 1201–1213. ↑133

[Séé1982] P. Séébold, *Morphismes itérés, mot de Morse et mot de Fibonacci*, C. R. Acad. Sci. Paris Sér. I Math. **295** (1982), no. 6, 439–441. ↑81

[Ser1985] C. Series, *The geometry of Markoff numbers*, Math. Intelligencer **7** (1985), no. 3, 20–29. ↑45

[Sim2004] J. Simpson, *Disjoint Beatty sequences*, Integers **4** (2004), A12. ↑8

[Šir1957] A. I. Širšov, *On some non-associative null-rings and algebraic algebras*, Mat. Sb. (N.S.) **41(83)** (1957), 381–394 (Russian). ↑133

[Šir1962] ———, *On the bases of free Lie algebras*, Algebra i Logika Sem. **1** (1962), no. 1, 14–19 (Russian). ↑42

[Smi1876] H. J. S. Smith, *Note on continued fractions*, Messenger of Mathematics **6** (1876), 1–14. ↑5, 44, 48, 49, 116

[Sta1997] R. P. Stanley, *Enumerative combinatorics*, Vol. 1, Cambridge Stud. Adv. Math., vol. 49, Cambridge Univ. Press, Cambridge, 1997. ↑117

[Sta1999] ———, *Enumerative combinatorics*, Vol. 2, Cambridge Stud. Adv. Math., vol. 62, Cambridge Univ. Press, Cambridge, 1999. ↑79

[SG2002] J. Stoye and D. Gusfield, *Simple and flexible detection of contiguous repeats using a suffix tree*, Theoret. Comput. Sci. **270** (2002), no. 1-2, 843–856. ↑123

[Thu1906] A. Thue, *Über unendliche Zeichenreihen*, Christiania Vidensk. Selsk. Skr. (1906), no. 7, 1–22; reprinted in T. Nagell, S. Selberg, and K. Thalberg (eds.), *Selected mathematical papers*, Universitetsforlaget, Oslo, 1977. ↑67, 97, 127

[Thu1912] ———, *Über die gegenseitige Lage gleicher Teile gewisser Zeichenreihen*, Christiania Vidensk. Selsk. Skr. (1912), no. 1, 1–67; reprinted in T. Nagell, S. Selberg, and K. Thalberg (eds.), *Selected mathematical papers*, Universitetsforlaget, Oslo, 1977. ↑67, 81

[Thu1977] ———, *Selected mathematical papers* (T. Nagell, S. Selberg, and K. Thalberg, eds.), Universitetsforlaget, Oslo, 1977. ↑67

[TS1995] J. Tromp and J. Shallit, *Subword complexity of a generalized Thue–Morse word*, Inform. Process. Lett. **54** (1995), no. 6, 313–316. ↑82

[Ukk1995] E. Ukkonen, *On-line construction of suffix trees*, Algorithmica **14** (1995), no. 3, 249–260. See also `http://www.allisons.org/ll/AlgDS/Tree/Suffix/`, where it is really online. ↑119

[Vie1978] X. G. Viennot, *Algèbres de Lie libres et monoïdes libres*, Lecture Notes in Math., vol. 691, Springer, Berlin, 1978. ↑42

[Wei1973] P. Weiner, *Linear pattern matching algorithms*, 14th Annual IEEE Symposium on Switching and Automata Theory (Iowa City, IA, 1973), IEEE Comput. Soc., Northridge, CA, 1973, pp. 1–11. ↑119

[Wel1984] T. A. Welch, *A technique for high-performance data compression*, Computer **17** (1984), 8–19. ↑114

[WW1994] Z. X. Wen and Z. Y. Wen, *Local isomorphisms of invertible substitutions*, C. R. Acad. Sci. Paris Sér. I Math. **318** (1994), no. 4, 299–304. ↑14, 37

[Wri1959] E. M. Wright, *Prouhet's* 1851 *solution of the Tarry–Escott problem of* 1910, Amer. Math. Monthly **66** (1959), 199–201. ↑71, 72

[Zim1982] A. I. Zimin, *Blocking sets of terms*, Mat. Sb. (N.S.) **119(161)** (1982), no. 3, 363–375 (Russian). ↑127, 131, 132

Index

Key words and phrases from the text appear in lexicographic order, as usual. Mathematical symbols appear in the order in which they occured within the text. Mixtures of symbols and text, such as *k-avoidable pattern*, are likely to be found among the key words (omitting the symbols), e.g., under *avoidable pattern*.